Kay Junge, Kirill Postoutenko (eds.)
Asymmetrical Concepts after Reinhart Koselleck

Histoire | Volume 20

Kay Junge, Kirill Postoutenko (eds.)

Asymmetrical Concepts after Reinhart Koselleck

Historical Semantics and Beyond

[transcript]

Gedruckt mit freundlicher Unterstützung des Excellenzcluster »Kulturelle Grundlagen von Intregration« (Universität Konstanz)

Bibliographic information published by the Deutsche Nationalbibliothek
The Deutsche Nationalbibliothek lists this publication in the Deutsche Nationalbibliografie; detailed bibliographic data are available in the Internet at http://dnb.d-nb.de

Cover layout: Kordula Röckenhaus, Bielefeld
Proofread and Typeset by Kay Junge
Printed by Majuskel Medienproduktion GmbH, Wetzlar
ISBN 978-3-8376-1589-0

Contents

Acknowledgments

This volume is based on the proceedings of the workshop *35 Years After Reinhart Koselleck: Asymmetrical Concepts in Politics, Language and Society*, which took place at the University of Constance on June 3-5, 2010. Whereas a conference grant of the German Academic Exchange Service (DAAD) made the event possible in the first place, the publication of proceedings was assisted by generous subsidies provided by the Center of Excellence *Cultural Foundations of Integration* (University of Constance) and the Alexander von Humboldt-Foundation (Bonn, Germany). Last but not least, the work-study team attached to the project (Matthias Buderbach, Thorn Kray, Francis Le Maitre, and especially Nils Meise) made editing a lot more enjoyable than it usually is.

Kay Junge,
Kirill Postoutenko

Self-Concepts, Counter-Concepts, Asymmetrical Counter-Concepts

Some Aspects of a Multi-Faceted Agenda

KAY JUNGE

> "[...] when the Spaniards asked the name of that place, [...] one of the savages answered 'Jucatan', which is 'What ask you?' or 'What you say?'" (Walter Raleigh 1614).

I A DISCLAIMER

In this introductory essay, the questions of what concepts are, how they refer to real, possible or fictive things and events by means of denoting, labeling, designating, determining, constructing, evoking, portraying, or identifying them, as well as how concepts absorb uncertainty, will only be touched in passing. Other questions which will not be extensively addressed here could be summarized as follows: How are words, sayings, phrases, ideas or concepts invented or made up, come into being, and made to be understood? How are they learned—by supervised problem solving, by the imitation of verbal behavior or by also figuring out the possible intentions that might orient such behavior? How could they intentionally or unknowingly be misread? How are they put into circulation? How do they get transmitted, bequeathed and inherited, reproduced or replicated? Why should they polarize and become contested? How do verbal signs get trimmed up or embedded, specified or generalized, or simply become ambiguous, how do they become conventionalized or fossilized? Last but not least, how do some elements of social communication manage to jump to key positions and sneak into larger traditions, while others might eventually turn out to be redundant and slip into irrelevance?

In fact, sociology, which often claims strong interest in culture, should be a proper place for asking and answering these questions. After all, humans have managed to protect themselves, survive and spread all over the globe by relying on cultural artifacts, concepts, instruments and institutions, which themselves are subject to the processes of selection and adaptation eagerly studied by sociologists. However, as sociology lacks intellectual resources for tackling the processes of individual learning and cultural evolution and does not even seem to recognize this deficit, it has tacitly delegated these problems to other disciplines. Being well aware that one does not live by genes alone, sociologists still appear to treat humans as cultural dopes incapable of possessing concepts in the first place. Moreover, they are often close to the assumption that human vocabularies determine human behavior in much the same way as genes regulate activities of other animals. As a result, it seems like sociology has little to offer when it comes to dealing explicitly with the smooth but intellectually very demanding inferential processes of communication and the communicative commitments they impute, with our mental bookkeeping and conversational haggling, our pretentious imputations and clever alibis, our strategic self-revelations and silencing tactics, and the intricate semantic architecture that gives asymmetric counter-concepts their peculiar flavor.

Reinhart Koselleck, who coined the very term *asymmetrische Gegenbegriffe* some 35 years ago, illustrated the architecture of asymmetric counter-concepts mainly by going through three examples. (Koselleck 1985). His first example was the opposition of Hellenes and Barbarians. This distinction owed its robustness to the fact, that what it referred to was also separated in space. The distance between them allowed Hellenes to look down at allegedly speechless Barbarians, and it was only after they had found a label for the Barbarians that they sought one for themselves: discriminating against others came first. This conceptual pair became dubious and contested when its territorial foundations lost their strength. The second pair of counter-concepts Koselleck discussed was the one of Christians and Heathens.

It did build upon the first distinction but also added a strong temporal connotations to the original, and thereby incited Christians to a more dynamic and active relation vis-a-vis others. The Christians acted as missionaries, and as a consequence they had to make a further distinction between brothers and potential converts (i.e. heathens proper), and those damned to perish (i.e. heretics), thereby adding a spiritual dimension to the distinction. The last asymmetrical binary analyzed by Koselleck was the pair Übermensch *vs.* Untermensch, which quickly acquired racist connotations but could also be, according to Koselleck, traced back to the Enlightenment discourse, legitimizing the decapitation of the king and others, and was also constitutive for the totalitarian discourses, purges and cleansings of the 20th century. Here the one side simply and brutally defines its counterpart as non-human in the first place, building up on the religiously driven negation of the other hinted at before.

At first sight, the conceptual asymmetries mentioned above articulate the value-judgement that apparently comes naturally, relying on a preference of self to other, or us to them. But as Koselleck's readers know, at a second glance things turn out to be more complicated, intricate and interwoven with one another. However, as the essays collected in this volume make clear, the complications that will surface also depend on the disciplinary framework applied, and there are quite a number of ways to derive, frame and embed counter-concepts. It is in bringing these different disciplinary perspectives together that we hope to make a difference in the current research landscape. One could in principle begin with answering the more elementary questions posed above, but, as has been already indicated, there are no ready answers to them. What could surely help is an even more ambitious interdisciplinary endeavor than the present volume, with psychology and cognitive science looming large—but that is, for now, a thing of the future. Hence we will have to start with a more limited, less elementary and also more abstract set of questions. First, we will ask why historians, social scientists and linguists should be particularly interested in cultural self-images, in self-concepts, self-fashioning and self-descriptions of particular times and peoples, specific encounters, procedures, gatherings and institutions. Next, we will address the problem of anachronistically applying today's concepts to other historical settings. In the remaining three sections we want to discuss the possible architectures of counter-concepts, both, symmetrical and asymmetrical, drawing mainly on the conceptual specifications and definitional procedures Carl Schmitt applied or touched on in his booklet on *The Concept of the Political* whose first edition appeared in 1927 (Schmitt 1996) and will eventually discuss three more varieties of instituting asymmetries from a list that is still open and should invite the reader to think of still others not tackled here. Reinhart Koselleck had close relations with Carl Schmitt while writing his dissertation, and took from Schmitt much of his inspiration for the idea of asymmetrical counter-concepts. This connection could therefore be a fitting point for introducing our book.

II SELF-CONCEPTS, THEIR INTRICACIES AND VAIN FIXATIONS

Right at the beginning of his *opus magnum* of 1651 Thomas Hobbes observes that man is both, the maker or artificer and of his institutions and the matter of these artificial, i.e. historically contingent, bodies. Of course, in their capacity as builders of their own social world, humans often fail and things turn out to be different from how they were intended—if they were intended at all. And this is why Hobbes, like arguably all political writers, endeavors to intervene into this process, trying to hold up a mirror to a man to enable him to better see who he is, to get a better grasp of his situation, and to transform society more to his liking. The metaphor of holding up a mirror implies an innocence and neutrality of such endeavors that could well be misleading. In fact, Hobbes is evidently not holding up a mirror to us, but is rather writing a book. But at least with reference to social matters there is no way out of the circle constituted by the fact that man is both the maker and the matter of his affairs. His words do not only serve as indicators of what is happening, but they must also be seen as factors in what is happening. Our language for the description of our social affairs does not only mirror an independent reality but partly constitutes it. Naming things and events, blaming others or ourselves for what happened, and claiming responsibilities (or demanding compensations) are actions that build upon each other and determine what happens next. Even being told the truth or gaining some self-knowledge will make a difference, change our understanding, and will perhaps make us act differently. The mirror can make visible things we cannot see otherwise—

most importantly, our own image and the people in our back. Holding up a mirror to someone and making him look at it is nothing else than an active intervention. Once social philosophizing or social research becomes known to the public, they inevitably change their object domain.

But the mirror metaphor might make us aware of something even more subtle than that. The mirror itself already seems to be doing something; we identify the mirror by its mirroring of the things in front of it. It seems to distinguish itself by what it does to its surrounding world that is by reflecting it. Of course, mirrors can be bent; they can also enlarge things, distort them beyond recognition or even make them disappear completely from view.[1] And in all this, again, they are a little bit like language in general and the specific concepts we use to depict the world, ourselves and others. Whenever such processes of self-reflection and self-description get lexicalized and condensed into special terminology, we speak of self-concepts. Inevitable abstracting from many details, but still being a part of what they describe or make us conceive, they cannot be but incomplete, just as a map of some territory, which is both used and covers some space within that territory, cannot map itself in all detail, or the mirror, that cannot mirror its backside.

Unlike artifacts such as mirrors, language, however, evolves without someone having some plan in mind. Because of that, it cannot be set up from its very beginning by explicit definitions and conventions,

1 Many if not most Marxists where not too perceptive of such possible distortions and tended to account for science and their own endeavors by the metaphor of the mirror-image, while drawing on the concept of ideology to account for the distortions of others. So called positivist did not do much better by conceiving of their own projects as science and all the rest as meaningless. In response to this split and apparent arrogance, some postmodern thinkers sought a way out by inflating the concept of ideology and social construction to cover all kinds of culturally shaped cognitive processes, but then talked themselves into denying the existence of mirrors, apparently even real ones, all together, thereby again missing something important, namely both the irritations that especially the apparently true images and propositions can trigger off, and the failures and falsifications that might be due to the fact, that we oriented ourselves to a mirage or some distortion in our mirror.

as these would have already presupposed an existing language. It has evolved and got shaped by our interests and circumstances, only to transform them in turn. However, human language has developed features, in particular our ability to say *No!* that make it quite different from the signaling systems we can already find among bacteria, plants and in the animal kingdom. Negation facilitates the specification of meanings, so to speak, on the spot and allows for the delimitation of referents by processes of triangulation, where e.g. you learn what I mean by "tulip" by pointing to all kinds of plants and flowers, or even other things, and I nod wherever I feel that you do not meet my expectations. It is only through communication and by negation that meanings get sufficiently fixed to allow for the coordination of action, and it is only through tacit confirmations and explicit negations that we arrive at some workable consensus. By relying on a specific kind of default logic, namely by proceeding on the assumption to understand each other and perhaps even agree with each other, as long as no one explicitly disagrees, our language allows us to communicate in a most efficient way (Horn 1989, especially chapter 3).

The default logic constitutive of human talk works by a preference for agreement. Agreeing goes unmarked, disagreeing needs to be marked. It is perhaps from here that the most elementary asymmetries of human communication and concept-formation take off. But, of course, we do not always arrive at a sufficiently wide ranging working consensus, and are oftentimes well aware of this. But still disagreements occasoinally come as a surprise and, if so, frequently require reorientation that is not easy to get at short notice. The maxim *I do, what you want me to do, as long as you do, what I want you to do* that tacitly guides our conversations, is threatened if an offer is rejected openly and this rejection is perceived as a threat to one's face. If such disagreement, in turn, gets rejected, a conflict is going to escalate, and it may end up dominating the whole situation. Furthermore, the conflict gets entrenched when parties begin to orient themselves to another complementary maxim that is only slightly different from the previous one: *I don't do what you want me to do, because you don't do what I want you to do*. Each party now attributes the fault to the other side and a preference for disagreement will give both of them a new stable orientation. As long as the parties cannot move themselves to agree on their disagreements, isolate them, simply forget about them, or at least pretend to have forgotten about them, conflict will prevail and often escalate by making the parties look out for further topics to fight about and by attempting to draw others into the conflict (for this communication-based view of social conflicts see Luhmann 1995: chapter 9).

But conflicts may also be sought on purpose. Discriminating against others for whatever reasons—and conflict, once started, will quickly provide good reasons—might serve as, or appear to be, a costly signal, precisely by cutting off options for making further contact with the discriminated. This is something not everyone can afford, and, once begun, it commits those who started it. There is no easy going back. It thereby might allow for more trust and better organization among those engaged in discrimination compared to those who are discriminated againt, and, should the latter already be ahead, it might be time to keep up with them (Posner 2000). Asymmetries might arise even in a completely homogeneous population. Adopting a maxim from Goffman, one could say: it is not man (and we may add: tribes, nations or cultures) and their moments, but rather moments and their men. But, however arbitrarily started, conflicts will soon have their own histories and parties will attach labels to each other and frame the situation in mutually incompatible ways.

Though few would agree with the Communist Manifesto of Marx and Engels in seeing the history of mankind as a history of class war, many conflicts are asymmetrical from their very beginnings. But there exist also others where social segments fight each other on a more or less equal footing. But then again, even though outsiders to the conflict might observe reciprocity between the parties, to the parties themselves the situations might appear inherently asymmetrical, with one's own position privileged and the other's denied recognition and respectability. In both cases, if the conflict is not resolved by the victory of one of the two parties, a third force intervention or mutual exhaustion, it will continue and become a part of, or even a constitutive element in each party's self-understanding.

As John Austin, following Herder and Humboldt, pointed out, language *embodies all the distinctions men have found worth drawing, and connections they have found worth making, in the lifetimes of many generations* (Austin 1970: 182). One needs not be a conservative to appreciate how much wisdom we inherit almost unthinkingly simply by growing up as language-using animals amidst traditions and all the extras that come with reading, writing and doing math with pencil and paper: nowadays college students may calculate faster than our ancient philosophers. The languages we have learned to speak direct our attention, enrich and differentiate our sensibilities, instincts and desires, and also unburden our thinking; words come to us easily. Ferdinand Saussure spoke of language in terms of system and structure. He took inspiration from the new for his time marginalist equilibrium economics, notably from Leon Walras, and thought that signs are constituted by relating a signifying sound to a signified thought or image (Piaget 1971). Signs, according to Saussure, get their specific meaning only in relation to other such signs formed in the same manner, just as the price of some product or service seems to be determined by the demand and supply for that specific product only in relation to the demand and supply for all other services and products. Just as a specific price vector balances supply and demand, a specific sign system matches the universe of signifiers with the universe of signified. The value indicated by some price is determined by all the other prices, just like the value of a certain sign is determined by the summary value of all others.

How could such balance and harmony come about? Leon Walras invented a central auctioneer, who collected all the information needed and then determined the price vector where supply and demand would meet. Trade was only allowed to begin after all prices had been determined. As for Ferdinand Saussure, he was not bothered much by the question of how the whole system might be set up in the first place: similar to equilibrium economists, structural linguistics downplayed the part played by each individual language user. In passing we should mention that another contemporary of Saussure, namely Emile Durkheim, while trying to establish sociology as an academic discipline, conceived society and its collective representations in similarly static terms. All three scholars departed from two complementary assumptions, namely that the system was already in balance and that the impact of a single individual was negligible. Thus equilibrium economists saw no need for bargaining, haggling or strategizing; structural linguists thought that speech (*parole*) did not affect language (*langue*), and Durkheim felt that individual performances were directed by our supposedly collective representations and close to incapable of subverting them.

But if everything is related to everything else, there is no leeway for change, and as price systems, sign systems and systems of collective representation incarnate perfect matches between signifiers and signified, there is no need for change, so to speak, from below. Sticking to these assumptions despite the fact that the world does not live up to these ideals, the disciples of Walras, Saussure and Durkheim were forced to take a normative stance: equilibrium economists polemicized against market-interventions, structural linguists fumed at the apparent prison-house of language and often elaborately embraced silence and the unspoken; and some Durkheimians tacitly began cultivating some reactionary ritualism, continuing by culturalist means the spiritual agenda of the French 19th century restoration thinkers that already inspired Durkheim in the first place.

There are interesting parallels to these intellectual developments, but we should not let them distract us much further, as it does not take much to see that there is less perfection and more fragmentation and complexity in our vocabulary and economy, than Saussure, Durkheim and Walras allowed for. Wherever there *is* a structure, there are structural holes, and these leave room for maneuvers involving trial and error; prices do change due to human intervention, and languages vary due to human invention or neglect. How do such admittedly blind variations get selected locally in order to possibly spread later on? And what are the localities or arenas they might take off from? Economists quickly identified oligopoly problems, problems of bargaining, and speculative bubbles. Sociologists began to scrutinize the interactional subtleties alongside the larger social consequences of labeling processes and socialization, and became more sensitive to the negotiation of social order at the shop floor level as they sought for explanations of inertia or even sudden collapse of larger institutional complexes. Linguists and discourse analysts began to differentiate between different areas, fields, discourses or sublanguages more or less isolated from each other and evolving independently, identified key-terms, analyzed contested concepts, processes of change and substitution, conventionalization and fossilization; more recently, they began to look at the microstructure of conversations and the local inventiveness of participants and engaged themself in the performance of deconstruction, or recommended an ironic subversion of the status quo.

III CONCEPTS AND THEIR HISTORICAL CONTEXTS: IDENTIFYING ANACHRONISMS AS A JANUS-FACED ENDEAVOR

Conceptual history and social history, the history of ideas and the history of events complement each other but can hardly be reduced to one another in either direction. A pure history of ideas, as it is often associated with the works of Arthur Lovejoy, is an impossible project for most modern researchers, though it might still be thought feasible where the complex of ideas under investigation consists in a cosmos closed onto itself which most probably presupposes that it is an entirely bookish tradition (see e.g. Grant 1981). Likewise a purely materialist social history disregarding the self-reflexivity of its object-domain has been abandoned by the majority of modern scholars. The materialist critique of ideology already tacitly indicates this failure and condenses it into a performative contradiction, unless economic determinism is taken to work in the long run only, with plenty of room for aberrations and false consciousness.

Our social expectations do not necessarily match our upcoming experiences, and equilibrium is not necessarily reached. As such, it almost never comes instantly, and it may well be that it never comes: as we stumble forward, we can be happy that we do not fall too often and too hard. Should the expectations we have and conceptually articulate fit the events occurring to us in a perfect fashion, we may characterize such a situation as being in equilibrium. It is only in such a case that we would have neither cause nor reason to revise our expectations. If our concepts give us a sufficiently tight grasp on reality, no one should have an incentive for revising commonly held notions, or at least no one should succeed in disturbing the established balance. Up until now, however, history has been more of a stumbling forward, and the equilibria we had were interim solutions at best, though in some cases holding for many generations. The vocabulary we inherited and employ to arrange our affairs may at times be outdated (which we usually notice only in retrospect), and it might easily mislead us in understanding our current situation. For the same reason our current vocabulary does not necessarily fit historical periods that have long passed. Social evolution and conceptual traditions and innovations need not be perfectly aligned.

In his book on the 18th Brumairc of Louis Bonaparte Marx took the mickey out of his hero by showing that Bonaparte, in order to install himself in the highest office, was forced to rely—for the lack of better alternatives—on a mode of self-fashioning already outdated at his time, just as Luther used to imitate Saint Paul and the protagonists of the French Revolution attempted to alternatively revive Roman republicanism and the cult of Roman emperors. Nowadays, research in historical semantics is at least in part inspired by driving out the inverse of such anachronistic misfits, criticizing the application of concepts adapted to our current situation to understand a past where different circumstances have prevailed. However, supplying concepts with a time index, as we will see shortly, can be a mixed blessing.

One of the principal works that lead to the introduction of historical semantics in German history curriculums was Otto Brunner's *Land and Lordship: Structures of Governance in Medieval Austria* (Brunner 1992), which first appeared in 1939. Brunner, who later, along with Werner Conze and Reinhart Koselleck, was one of the founders and editors of the path-breaking encyclopedia *Fundamental Concepts in History*, criticized in his book the prevalent discussion of late medieval polity in terms that only became common during the 19th century, i.e. in terms of the modern state, which resulted in missing and misconstruing certain central features of medieval Europe and portraying medieval polity and social order as a primitive anarchy. Instead, Brunner proposed historians to attempt understanding of their objects of study in terms prevailing at their times, and arrived as a result at a much more harmonious and orderly concept of late-medieval Austria to which we will come back later.

But Brunner's own terminology itself soon proved to have been anachronistic and slowly turned out of fashion after the war. Brunner, of course, went with the flow and, without giving up much, rephrased his former *Volksgeschichte* as a *structural history*—the term that since has become ubiquitous. Brunner was convinced that one should approach social reality justas Schmitt repeatedly argued by identifying its concrete order, not by relying on artificially set up terminology and positivist legal abstractions. This insistence to be as true as possible to the hard facts and the ways they relate to each other is surely commendable. But what could be recommended to a historian should not necessarily serve as guidance to a lawyer. After the Second World War broke out, Schmitt became increasingly concerned with the question of how the different political orders that we know of correlate with certain weapon technologies allowing them to occupy certain territories and control the sea or even the airspace. The modern nation-state, he felt, should already be seen as a thing of the past. Confronted with worldwide markets and global competitors, it could no longer control and navigate its own socio-economic life. For those like Schmitt and Brunner it was therefore only natural that the dawn of a new spatial order had to be articulated in new, tangible concepts. The supposedly mutual relation between protection and obedience that Brunner and Schmitt thought to be constitutive for political order, had to be reformulated with reference to larger territories that Nazi-Germany was just about to conquer. After its defeat, though, the problem, although still not very well articulated, remained. Today no one would deny that, while history has advanced beyond the confines of the nation-state, we still lack a proper vocabulary to handle this new situation. This brief inquiry into some of the contexts of Brunner's work was meant to raise our awareness of ambivalences and side effects, intended or not, that are inherent in every criticism of anachronisms.

In Britain, the misguiding and potentially blinding impact of the anachronistic use of concepts and the dangers stemming from neglecting their specific historical contexts, began to be reflected upon just about the same time, and latter set the stage for the arrival of the so-called *Cambridge School of Political Thought*. We would like to mention here only three authors that in quite different ways shaped this arrival, namely Robin George Collingwood, Lewis Bernstein Namier, and Peter Laslett. In his influential and much read autobiography the British philosopher, archeologist, and historian of Roman Britain Robin George Collingwood wondered whether Plato's Republic and Hobbes' Leviathan should be taken to represent just *two different theories of the same thing* (Collingwood 1939: 67). He answered in the negative: *Plato's Republic is an attempt at a theory of one thing; Hobbes's Leviathan an attempt at a theory of something else* (Collingwood 1939: 62). Plato and Hobbes wrote about different things, considered different things to be possible and desirable in terms of their social organization, and asked themselves different questions: indeed, the ancient polis was one thing and the absolutist state another. To conceptualize the Athens of Pericles according to the parameters of the modern or early modern state might therefore amount to no more than *logical bluff*, of which Collingwood accused his opponents. To *rethink the thoughts*, as Collingwood had it, of Plato or Hobbes, one needs to take off the conceptual blinkers of one's own time. Collingwood's scathing attack at the history of ideas, however, made comparisons across time and space look like a vain endeavor.

At about the same time the historian Lewis Bernstein Namier, having been attracted by the sociology of elites of Vilfredo Pareto whose lectures he had attended before emigrating to the United Kingdom and becoming a British subject, had been working on the composition of the Parliament of Great Britain and particularly on the individual careers and biographies of its members during the late 18th century. Namier's impact on the history of ideas is mainly negative, but that does not diminish his relevance. He convincingly refuted the idea, prevalent among most of his contemporaries, that 18th century politics was already the party politics where Tories and Whigs confronted each other. Instead, Namier showed, that local interests and universal corruption, (which he took to be a mark of *English freedom and independence*, since, as he pointed out *no one bribes, where he can bully*) could explain the careers and voting records of parliamentarians much better (Namier 1961). His dissolution of collective categories, ideals and ideas has become known as *namerization* and the verb *to namerize* has even been added to the Oxford English Dictionary. Reducing ideas to mere window-dressing hiding away the question of *cui bono*, he might have anticipated revisionist historiography long before it got its name. But he also made us much more alert to the fact that our concepts and what they refer to should not be trusted blindly and taken for granted. Battling against Namier, Herbert Butterfield has shown a way to do intellectual history without being a Whig.

Historian of the next generation, Peter Laslett discovered the library of John Locke. He was able to show that Locke's *Two Treaties on Government* could not have been written to legitimize the Glorious Revolution, as it was commonly assumed, but were already put on paper ten years before the event. This discovery was obviously difficult to top, but in the preface Laslett wrote to his 1960 edition of Locke, he also provided the impetus for the Cambridge School of Political Thought for which the names of John Pocock and Quentin Skinner stand in. Taking into account Collingwood's idealistic individualism, Namier's attempt to substitute local detail for conceptual generalizations, and Laslett's lucky discovery placing Locke's *Treaties* into a new context, one could have already expect the conceptual historians of the Cambridge School to pay closer attention to individual biographies and textual reconstructions, working at a greater distance from social history (let alone such notions as class) and being more averse to explicit generalizations and abstract conceptual schemes, then their German colleagues working in the field of historical semantics. And so it turned out to be. However, in retrospect, we can also identify a number of larger agendas in and political affinities and implications of their research: for Skinner, that would probably be the recovery of Roman rhetoric and the contextual reconfiguration of our notion of liberty, and in case of Pocock one should speak of saving from oblivion the classical republicanism in English and Anglo-American political thought with its stress on the active citizenship.

This difference of foci does not necessarily put the Cambridge scholars at odds with their German counterparts. Compared with 19th century attempts at the history of ideas, the research done since the Second World War has become much more specific. General accounts of the spirit of certain times and places seem to be a thing of the past, with the exception, perhaps, of books written for non-academic audiences. Among other things, this change is due to a methodological innovation. Both Koselleck's attempt to link the space of experience with the horizon of expectations and his observation that in the modern world historical knowledge ceased to be a guide for the future reworks Hegel's model of the origin and the future drifting apart as a constitutive factor of the modern age. In a similar vein, Arnold Toynbee's famous schema of challenge and response was invented in part to salvage whatever was still considered valuable in the older philosophies of history. However, it was Collingwood's metaphorical use of the question-answer scheme that furnished the history of ideas with the new precision needed to release it from the clutches of idealist historiography.

By relying on this schema, the Cambridge scholars could establish and maintain the distance needed not only from Leo Strauss's notion of a timeless and disembodied canon of political ideas, but also from the pure history of again quasi-Platonic eternal ideas for which Arthur O. Lovejoy was said to stand. Arnoldo Momigliano makes clear what was at stake here by associating Lovejoy with the polemics against Margoliouth, who had the reputation of believing in the existence of thirty Indo-European *Ur-jokes* from which all others were said to be derived, noting that Lovejoy had not considered the number of *Ur-ideas* to be much greater than that (Momigliano 1977: 6).

In the German scholarly discourse, the attempts to understand changes and transformations of ideas as responses to specific questions or challenges which the actors initiating such changes felt obliged, respectively, to answer or respond, was mainly associated with the work of Hans Georg Gadamer. Understanding something in this context means understanding it as an answer to a question. But it was Hans Blumenberg, who later on has refurbished intellectual history by innovatively applying to it this question-answer scheme. Questions that once have provoked certain answers might disappear from view, no longer be regarded as challenging or even be forgotten. But we might still have the answers at hand and therefore start seeking new problems to apply them to. If Christianity offers salvation, it's got some explaining to do: salvation from what? And so, to answer this question, one may have to invent eternal sin. It is not only that questions ask for answers, it is also that answers seek to be adopted by questions in order to remain intelligible or become intelligible again. The logic of question and answer must be conceived of as identifying a relation that is variable on both ends.[2] With this newly acquired flexibility Blumenberg could argue for the *Legitimacy of the Modern Age* (Blumenberg 1983), showing that modernity was not an illegitimate child of religion, and that those who denied its legitimacy, who thought its self-assertion and piecemeal progress ill founded, and asked instead for absolute foundations, foundations which only theology could offer, suffered from an intellectual hangover of a bypassed age. Blumenberg has demonstrated the secularization thesis to be an anachronism, and argued that the burdening canon of questions that the theologically inspired anti-modernists wanted to be answered and which made them willy-nilly turn into reactionaries, could be handled fare more flexibly and—at least in part—be dispensed with.

Of course, not everybody agreed: Koselleck, for instance, did not, although he refrained from either arguing with Blumenberg or defending his opponents. Those asymmetrical concepts in which he was in-

2 It should be mentioned in passing, that what Blumenberg did to Carl Schmitt, Erik Voegelin and Karl Löwith, Niklas Luhmann did to Talcott Parsons by inserting contingency into the structuralist premises of the latter' approach.

terested most seemed all to have Christian pedigree, and insisting on their understanding as secularized variants of older religious notions was meant to baffle those who, like Cold War protagonists or the emerging New Left, according to Koselleck kept sticking to them in various ways, knowingly or (more often) unknowingly. Discriminating against the notions they held to be products of secularization, of course, may by itself be seen as an application of an asymmetrical counter-concepts.

Indeed, some historians, or for that matter, cultural sociologists may be seen to employ asymmetrical counter-concepts. But maybe they shouldn't. Among the most conspicuous conceptual pairs of questionable value one could name the popular distinction between myth and reason surviving under various, mostly structuralist, culturological or psychoanalytic guises. Its asymmetry might surface in different ways. The most obvious one being this: While we in our classrooms discuss, compare and reflect on different mythical constructions, we often at the same time assume that our sources were incapable of doing so, blindly believing in their narratives. Paul Veyne and, more elaborately, Hans Blumenberg have questioned such assumptions: myths, as long as they have not been transformed into dogma by political institutions, have always been contested, discussed, amended, improved, mixed with other accounts and in parts forgotten. But we don't know for sure. The real surprise, brought to us by the ethnographical research in oral cultures is that certain epic accounts survive at all in an identifiable fashion. What we might be inclined to classify as different archetypes, Blumenberg sees as products of selection brought about by thousands of years of storytelling, i.e. as the results of optimization (Blumenberg 1985, see in particular the again excellent introduction by Wallace). It is not some archetypical imagination that shaped our mythic pantheon, but a *Darwinism of words*, as Blumenberg puts it, a selection process for which the reactions of the audience were central, and that accounts for their shape, complexity and eventually achieved resonance. This shift in focus from content to the medium and the institutions of transmission was got further support from research in anthropology, particularly the works of such contemporary anthropologists as Jack Goody and Fredrik Barth. Goody, working on different recordings of the Ladagaa myth of the Bagre, found out that they were quite distinct in what they related, and suggested therefore to put more stress on the relevance of creative variation and imagination in predominantly oral cultures, and be more skeptical about the assumed constants in the structuralist and functionalist accounts of myth (Goody 2010). Likewise, Fredrik Barth (1990) opined that the oral traditions of knowledge prevalent in those few simple societies for which we have reliable records, and in particularly the ones that rely on secrecy should be seen as highly distributed, fragmented, diversified, fluid and fragile systems. A closer look at the mechanics of communication and its carriers, their social organization and the processes of competition among them, tells us more about knowledge (re)generated, varied and passed to the next generation than any thick description of archetypes, in which constitutive role modern tourists are made to believe (Hobart 2000). Nevertheless, the emphasis on the means of communication and institutional reproduction will not necessarily endanger the counter-concept of culture in general: rather, it will allow for scrutinizing and perhaps putting into question its explanatory ubiquity.

IV CONCEPTS AND THEIR COUNTER-CONCEPTS: RATIFICATION AND ASYMMETRY

The approach promoted in this preface is to see concepts coming in pairs from their very beginning (even if the specific parts of each distinction are not always clearly marked from the start), rather than treating them as isolated phenomena which are paired or otherwise contextualized only by researchers if the need arises.[3] It is pairs of concepts giving profile to each other that we will focus on. Their mutuality, dialectics or (to put it in Buddhist terms) dependent co-origination has been observed and discussed for a very long time: one could mention, for instance, the fragments of pre-Socratic philosophers in the West, or the treaties of Nagarjuna in the East. The interest in these matters regained a new intensity among German idealists, only to fall again into oblivion after their heyday. (Perhaps such poets and philosophers as William Blake and Johann Friedrich Herbart, or Ralph Waldo Emerson with his nice pairings of essays on love and on friendship or on heroism and on prudence respectively, deserve to be mentioned here,

3 An alternative to this approach is explored by Kirill Postoutenko (see articles below).

too). A more focused interest in relational concept generation was only articulated in the last century. Especially students of law, with their more pragmatic approach, began reconsidering some of the fundamental legal concepts by identifying and relating them to their opposites and correlatives. Starting with Wesley Newcomb Hohfeld at Yale already before the First World War and ending with Norberto Bobbio's work pursued until the end of the 20th century, Carl Schmitt entering the scene after the First World War was not alone in his attempts at counterbalanced concept formation, and neither was Koselleck picking up his work.

As has been pointed out above, not everything in our semantic universe depends on everything else: most of our words and concepts are interrelated only in very, very indirect ways, having almost no influence on each other. Although learning all words and concepts from a dictionary would eventually make us run in circles, the entries we find in our dictionaries lay out very specific paths. Some words are just different and do not affect each other very much, whereas others' are distinct, their meanings being distinguished by their implicit or explicit connections to their complements. They have relational meanings, i.e. each gets its specific meaning in contradistinction to other terms: brother is related to sister and parents, aunt related to uncle and nephew, employer to employee etc. These words do not function as names for entities, but place things in contexts and in relations to each other. Some such words mark a relative, though perhaps calibrated, magnitude or intensity on some specific scale, for example, cold, warm, and hot; none, few, some, many, and all; or unhappy, not happy and happy. Other types of words have each a number of correlatives, like our kinship vocabulary, and some also have historically specific antonyms. Liberal is opposed to conservative, idealism to materialism, myth to reason, democracy to dictatorship, public to private, right to duty, and liability to immunity.[4] Although history of concepts traditionally deals

4 The difference between graded or scalable differences and binary categorical opposites has been elaborated first by Charles K. Ogden (1967), but the analysis of their use has greatly advanced since then (see the article of Lynne Murphy and Roberta Piazza in the present volume), gaining very much in depth and sophistication from the appearance of post-Griceian pragmatics, which is now part and parcel of every textbook in the field.

with the latter group only, the essays in this book take into account other groups, too.

Words and concepts may often invoke certain situations or images, but they become specific only by being concatenated with other more abstract, more specific or plainly opposite concepts. It is only by their specific concatenations that they might be seen as mirroring reality. It is not reality that rejects the concepts we hold. Rather, this rejection takes place when, being offered other concepts that contradict the ones we hold, this makes us abandon the latter. It is new concepts that might make us revise our conceptual networks, and these are usually offered to us by others that reject our views. It is by such conflicts and competitions in front of different audiences that our conceptual frameworks or networks of beliefs might eventually mirror reality in more or less robust and viable ways, acceptable at least for those who are close to us or socially relevant for us, or whom we cannot avoid. Socially enforced reduction of dissonance is certainly central in the selection of concepts. Having finite computing capacities, we try hard not to contradict ourselves and avoid being driven into corners or made fools by some Dutch book gambit exploiting our inconsistent, i.e. non-transitive preferences that economic textbooks warn about. However, dissonances need not always be faced, and some contradictions can be circumvented, concealed or tabooed instead of being addressed. Asymmetric concepts gain stability by discriminating against those who might be most inclined to contradict them.

When one antonym is openly or tacitly substituted for another, the meaning of the concept it has formerly given profile to will also change. If we contrast the public to the private, the second word in this pair of antonyms has some legitimacy for us—at least today, in the 21st century. However, the situation would be quite different if we contrast open access to secrecy or sectarianism. If an ancient philosopher wondered about what was specific to a man, he would perhaps have sought for criteria that made humans different from other animals. Alternatively, if a late medieval mystic were confronted with the same question, he might have pointed out, I fancy, at his differences from God and angels. Finally, if the question were posed to us today, we might come forward with a list of features telling men from women. More examples may be helpful in precising the issue. Let us take a look at three apparently well entrenched and often contested concepts: skepticism, competition and self-interest. What these concepts mean depends very much on the discourse and the implied antonyms that lend them their specific profiles. If skepticism is contrasted with enthusiasm, it will appear in a different light than when contrasted to wisdom or common sense. Competition seems preferable to monopolization but might only be recommended with reservation as an alternative to solidarity and brotherly love, and it may not be the right ideology for attracting blood donors. In its turn, self-interest looks perhaps narrow-minded when opposed to public-spiritedness but appears quite healthy when compared with self-hatred, self-mortification, glory, conspicuous consumption, status display or paranoia. Stephen Holmes (1993) observes that the anti-liberal rhetoric of Carl Schmitt, Leo Strauss and many others is based on the tacit substitution of antonyms which makes any liberal look like our stereotypical liberals do today. However, one could also argue that this falsification of the opponent's vocabulary makes visible the latter's conceptual blind spots. Our concepts gain in profile, when we make their counter-concepts and correlatives explicit, and they get subverted when we replace the implicitly held antonyms by others.

Most of us, I assume, would certainly prefer a situation when we had only to distinguish between friends and strangers, or friends and non-friends. To that Schmitt would respond that situations we find ourselves in are not always like this, and sticking to the harmoniously sounding way of partitioning your social world will only force you to label your enemies by some other word and classify them as criminals, outlaws, rogues, terrorists, or psychopaths. To get rid of them, or at least hold them at bay, you will not, of course, be obliged to engage in war. You will give your engagements a non-military label, even if others are likely to perceive it as a cynical euphemism. The claims of "bringing peace", "protecting human rights", or "preventing crimes against humanity" may easily be suspected to serve as a blank check for legitimizing military intervention. But these formulas should not be debunked as mere pretexts: one does not necessarily have to cast suspicion on the motives of such interventions, which may well be sincere and as such deserve our moral approval. It would be more pertinent, however, to focus on the specific rhetorical gambit employed in such cases, i.e. substitution of antonyms. Those who claim to know only friends, are said to treat those they don't know as criminals or psychopaths, and will thereby deny them the right to speak on an equal footing.[5] This in turn, according to Schmitt will put the pacifist (and later the reeducation engineer) at odds with his own self-image, making him more, rather than less, war-prone, since one can only make peace with an enemy, but not with a criminal or a psychopath. This all would have sounded more convincing for us today, and Schmitt's reluctance to embrace the global pedagogical ambitions of the newly established League of Nations at the cost of the old, if idealized standards of international law could have earned more sympathy if Schmitt did not become a Nazi shortly after having put his finger on the pacifist's blindspot.

The asymmetries that concern us here will be called into being either when the statuses of the participants of a certain social setting begin to differ considerably, or when their statuses and interests, or what they take to be such, begin to appear to be mutually incompatible to them. Our institutions and professions are built upon the first type of asymmetry. The asymmetry between government and administration might serve as an example, or the asymmetries that hold between doctor and patient, teacher and student, employer and employee. Sociologists call such relations *complementary* in contrast to the *reciprocal* relations connecting comrades, or teachers, or pupils, or neighbors, or friends respectively. Between reciprocally related equals, however, the probability of conflict seems higher, which constitute the second type of asymmetry (and there might, of course, also be mixed types). As the outcome of seemingly unimportant squabbles might be used as a precedent for deciding more serious conflicts in the future, both sides are inclined to invest some energy even upon trifles (Gould 2003). Actively engaging in such seemingly petty quarrels, advancing their partisan causes, legitimizing own positions with additional reasoning and appeals to common values (as if the other party does not embrace them), the parties involved find it increasingly difficult to treat each other on an equal footing, creating mutually incompatible, asymmetrically accentuated definitions of the situation.

Both reciprocal and complementary relations get an additional twist when they are not ratified by all parties involved. A few examples will suffice: your neighbor treats you as a stranger; your govern-

5 The term *footing* referring to the status of social participation in encounters, social gatherings or larger, perhaps even anonymous context is borrowed from Erving Goffman (1981).

ment holds you for a spy; a policeman pretends to be your friend; your husband demands money for doing the dishes; someone, having accidentally overheard an intimate conversations, interferes with comments on the issue; overall, those who are denied the right to speak because they lack logos, brain, culture, proper genes, purity, grab the floor and, worse, make themselves clear. In all these and similar cases the participation status of the parties involved is called into question and becomes contested. To be sure, these illustrations are mostly idiosyncratic, but the last example refers to sets of situations which are homogeneous and general enough to result in consistent labeling of the factions involved: those branded by one party as terrorists could be defended by the other as patriots or freedom fighters, and the contemptuous reference to crowds could be countered with a praise to people. The problem is that someone labeled terrorist, or relegated to the mob, would not be very much inclined to ratify this labeling. The use of asymmetrical counter-concepts entails addressing others in a disparaging way, and the party referred to, despite its awareness of being addressed, does not consider itself properly recognized by this reference. This disapproval, in turn, is likely to result in antonymic substitution of the kind described above. The interpellation, to use Althusser's term, might turn out to be not felicitous. It is here that counterconcepts become asymmetric.

Similar to most Germans of his time and some European contemporaries (John Maynard Keynes is most frequently mentioned in this context), Carl Schmitt himself was disinclined to ratify the new asymmetries constitutive of the newly established international order after the First World War. Instead of making peace in accordance with the classical *Ius Publicum Europaeum*, the treaty of Versailles had established a new discourse, new procedures, new hierarchies, and a new international order crowned by the *League of Nations*. The feeling that Germany was denied to speak on an equal footing with the other nations at Versailles and instead was being treated as guilty for having apparently started the war, and consequently forced to deliver reparation payments was what drove early Schmitt's polemics.

V The Distinction between Friend and Enemy and its Possible Asymmetrical Twists

Our experiences get structured by distinguishing what is present from what is absent, what is past from what might come, but also by differentiating between here and there. As soon as people recognize they live in company or begin voluntarily to form a group, the distinction between here and there gets transformed into the opposition of inside and outside, which to some extent may free itself from specific spatial connotations. There can be many such groupings and institutions, and there can also be multiple memberships, membership in a political community being one of them. Politics used to be considered the most comprehensive system, as long as society could be described as a territorially bound political unit. Modern society, however, has outgrown these notions and can no longer be grasped in that way. We lack adequate concepts good for our present situation in the way Aristotle's account was usable for Greek polices, Hobbes's for the 17th century, and perhaps the Federalist's Papers at the end of the 18th century.

In order to influence the situations we are in, we have to make ourselves predictable to others, as only this in turn allows us to anticipate their behavior and possibly affect the course of action. This is true not only for individuals but also for groups and institutions. However, self-presentation goes always along with some hidden or open altercasting, to borrow a term from the symbolic interactionist tradition (Hewitt 1997: 153-155). Such altercasting might rely on certain binaries, and cultural sociologists—at least those with a strong Durkheimian or structuralist bent—have taken such binaries more or less for granted and consider them to be constitutive for any moral order (one of the best books in this field is Smith 2005). Still, it is not unreasonable to suppose that such vocabularies may possibly develop from more elementary constellations. Pursuing this hypothesis may shed light at the rhetorical and communicative procedures that may account for their often asymmetrical twists, which make mutual recognition difficult or impossible, and often thwart all attempts at respect right from the start.

The American pragmatist George Herbert Mead argued that we usually learn to form an image of ourselves by observing how others react to our presence and our behavior, by attempting to observe ourselves from their point of view, and, finally, by appropriating their apparent image of us. Already Aristotle argued in his *Nichomachian Ethics* that friendship is based on our capacity to imagine ourselves from the others point of view. Carl Schmitt claimed that our enemies question us in an even more basic and existential way, and this being questioned leaves but very few options: what one side gains, the other may or even should loose and the best thing one can do or prepare for is to avert the worst. While games of mixed motives where there is some incentive for competition but also a benefit to cooperation usually have many equilibria, zero-sum games, illustrating pure conflict, admittedly, have only one, and the prospects of Schmitt's constellation seem in this light rather bleak. However, Marshall Sahlins has shown that there well may be a gradual difference between the positive reciprocity between friends and negative reciprocity holding between enemies—if the latter parties meet at all (Sahlins 1972). If the political, as Schmitt proposed, stands for the intensity of association and dissociation of human beings, whose lives may be put at risk by confronting each other as collectivities, than one should be able to locate this intensity on Sahlins' sliding scale (for more sophisticated and adequate multidimensional mappings of different types of simple two-person-games see: Kelley/Thibaut 1978: 75-110). The only difference would be that Schmitt more or less since the first publication of his essay did no longer conceive of the relation between friend and enemy in predominantly territorial terms.[6]

Schmitt conceived of the international order in Hobbesian terms and Hobbes warned that even in a world of states, the problem of order, that we today associate with his name, would not disappear but rather resurface, so to speak, on a higher level as these states would now confront each other, much like individuals confront each other in the state of nature. Even today many philosophers, sociologists and historians (e.g. Blom 2001) hold the so-called *prisoner's dilemma* to be a fitting illustration for this state of nature, although most game theorists doubt that this particular game is sufficiently complex to represent the problem identified by Hobbes. One may also wonder if there

---

6 Heinrich Meier attributed this shift to the influence that Leo Strauss has had on Schmitt. However, Günter Maschke, Schmitt's intellectual bodyguard and occasional editor who is usually better read in these matters than most academic Schmitt-scholars, pointed at some earlier essays by Schmitt indicating that this change occurred before he came to know Leo Strauss.

is here perhaps a category mistake: if it is the life of the players that is at stake in the Hobbesian state of nature, the ranking of the games' outcomes may be the wrong means to capture this. Can the existence of the players to whom we attribute such preference rankings be also a possible outcome? Maybe it can—to some extent we risk our lives in many things we do.

With prisoner's dilemma awaiting them, the players, to put it mildly, might be reluctant to show up. But if they have to, they might be inclined to fool one another into believing that they will cooperate. However, in a single-shot game where your behavior today makes no difference tomorrow, the incentive structure of the game suggests that both players should choose their dominant strategy and not cooperate. Whatever your opponent does, it is always better not to cooperate. The game has only one equilibrium, and no cheap talk will make a difference for those sufficiently familiar with its rules. As the parties are not capable to make mutual commitments and exercise influence upon each other, they will not cooperate. This game does not get us very fare as the parties involved will remain trapped in the state of nature.

Hobbes, however, identified three different causes constitutive for entrapping people in a state of nature. People might invade on each other for gain. This to some extant is perhaps captured by the prisoner's dilemma game. But anticipating their potential lot, they might also be inclined to invade for their own safety, attempting some preemptive strike. To capture this, the so called assurance game might be considered. This game is more interesting because it has two equilibria and because cheap talk and trust building can be of relevance in choosing one equilibrium over another. Consequently, this game also opens possibilities of asymmetric altercasting: each party can claim that it could be more cooperative, if only the other party were more reliable or, for that matter, less paranoid.

The third reason for invadingeach other, the one that Hobbes thought perhaps to be the most pernicious, is the desire for glory. This desire builds on and presupposes the other two motives. We might care for glory and attempt to gain a reputation for desiring glory to keep the others at bay. This to some extent might be captured by the socalled game of chicken. Facing a game of chicken should predispose each player to making a reasonably binding commitment before the other does, and to dissolve the commitments the others might already have tried to establish. Under such circumstances, trying to altercast the other as a weakling and to present oneself as the master of the situation might be a good strategy. Even presenting yourself as a risk-loving drunk hardly capable of changing your mind could be a strategy worth considering. The outcome would only be a disaster if both parties feel sufficiently committed to their glorious postures. The game comes close to the first round of Hegel's master-slave scenario: the one who risks his life becomes the master, whereas the one who fears to do so ends up being a slave. The chicken game has two equilibria, which are both asymmetric: either you play hawk while I play dove, or vice versa. Any other outcome will be regretted by both parties. The incentive structure of this game invites players to engage in impression management and bluffing and its two asymmetric solutions will make the contestants eventually to have to come to terms with their unequal footing reflected by the game's outcome, should they have arrived at one of its equilibria and not have killed each other or both have been frightened away from playing tough.

One might easily doubt that such seemingly naturalistic reconstructions and reductions will get us very far. But game theorists think of games in more complex ways than a cultural relativist or a social constructivist might suspect, and would hardly consider the three games mentioned above as more than simple toy games. But perhaps even more important is this: unlike 19th century utilitarians, the economists of today do no longer simply assume specific preferences. Rather, they look for such preferences to be revealed in the choices people actually make: it is only then that they would attempt construing some utility function that does match the choices made and that will possibly allow to forecast people's behavior in similar situations in the future. However, it seems doubtful whether such preferences can always be determined apart from their social settings, where, as Paul A. Samuelson has pointed out almost half a century ago (Samuelson 1966), they are often concealed and people sent false signals for strategic reasons. But if preferences are not determined empirically and in isolation and utility functions are instead just more or less made up, we risk arguing in circles, as the strategic reasons for the choices we may observe cannot empirically be distinguished from the preferences we assume to hold.

But here we can detect a happy coincidence between some of the methodological troubles economists, sociologists, and anthropologists find themselves in, and that should hopefully lead to greater understanding on all sides. Anthropology and sociology today agree that one should not simply assume that people hold certain values, share a certain culture and are members of certain collectivities. Instead, the researcher must ascertain under what specific circumstances the people in question exhibit behavior that could justify such attributions, being at the same time fully aware that those under observation often engage in pretty similar imputations. The situation gets even more tangled up by the problem of framing, which also has been extensively dealt with by experimental and behavioral economists. Just like everybody else, the objects of research may easily talk themselves into something they have never really considered before, and in the end feel even committed by such talk. They might perhaps even try to appropriate our academic categories, should this turn out to be to their advantage. Finding out, what people really are and what they really mean might easily turn into a self-fulfilling prophecy. Cultural sociology and cultural anthropology have therefore been criticized by more reflective empirically minded scholars for being *an enterprise that systematically confuses data with analysis and words with people* (Moerman 1974).[7] Simply by becoming aware of being observed, identity might be transformed into an achievement. Instead of being the referent of our narrative reconstructions, it gets promoted to the status of agency. But maybe this has always been the case and getting out of this loop by separating situations from the vocabularies they may generate is really close to impossible—or at least very difficult to achieve. Experimental semantics is still a thing of the future.

Before going back to Schmitt, let us recapitulate some of the ideas outlined above. Counter-concepts do not have to be asymmetrical. Valley and mountain, guest and host, or even left and right are counter-concepts having relational meanings, but are not asymmetrical. Counter-concepts become asymmetrical only when they distinguish between speakers, types of people, their groups or social roles, and when their use is not endorsed across the partitions they make. Counter-concepts can be asymmetric in a trivial, though still not very well

7 Worse still, as some critiques have pointed out, it might turn into an enterprise that by its culturalist readings and constructions, collapsing everything into patterns of culture, might be likely to supply further and more respectably looking semantic weapons for ethnic or nationalist conflicts or even for the clashes of civilizations, where reference to race has meanwhile luckily been found obsolet and publicly been banned (Kuper 2000).

understood sense where deictic references are at stake, i.e. where the conceptual binary allows you to mark your position in a relational way and this mark serves as an index for yourself. You can never be on the other side of the street, because once you arrive there, the other side has already moved across the road. In addition to these unquestionable counter-conceptual asymmetries, there are others where the existence of those they divide is at stake: when two groups claim the same territory, when two drowning men fight for the last place left on a raft, when only one can be the prizewinner. It is in anticipation of conflicts between mine and yours, me and you, us and them that the troubles, invited by some counter-concepts and hidden by others, begin.

Carl Schmitt's fame is not due to his definitions, but perhaps more to his aphorisms. For defining something, you first need to specify a *genus proximum*, a more general term, and then you have to propose a *differentia specifica* and identify the feature that makes what you want to define special and distinct. This, at least, is the classical way to arrive at precise definitions. Such definitions might not always proof to be very useful—a classic example here would be the definition of man as being a featherless biped. But as long as the qualities distinguished and marked by our terms can be easily observed and identified, such definitions are generally reliable. For all their inspiration, Carl Schmitt's ways of introducing concepts were often much less precise than the definition of human being just given. Due to the manner in which he introduced and placed his concepts they often gave the situation specified by them an asymmetrical twist right from the beginning, though, maybe this only proofs Schmitt's perceptiveness and sense of reality.

According to Schmitt (1996: 29-37) it is the degree of intensity of conflict that constitutes the political by identifying an enemy, and it is this intensity that makes my political enemy different from a mere market rival or a discussant in parliamentary debate. In Schmitt's terms, the fact that someone's life is at stake is an existential issue, whereas enemy is a *scalar*, or gradable, term. As a conflict breaks out or is expected to break out, anticipation and from then on escalation become attractive options., The chances of such escalation go up whenever the clear distinction between war and peace is not obtainable, so that the conflicting parties, in Thomas Schelling's words, lack a focal point needed for their mutual orientation. For Schmitt, a case in point was the promotion of economic sanctions to a legitimate instrument of international politics. However, appearances are deceiving, and social conflicts are seldom absolute, even when contestants are inclined to use absolutist rhetoric, if only to sell it to their constituencies. Poison gas was extensively used in the First World War, but none of the contestants in the Second World War dared to use it. The atomic bombs were thrown on Hiroshima and Nagasaki but never used in war after, and openly considering their employment became a taboo or a sort of a hidden vice asymmetrically ascribed by Cold War enemies to each other. Despite appearances sustained by our blinding existentialist vocabulary there often still is a premium upon cooperation. The main challenge is to find out how probabilities and expectations, attention attracting focal points, and concepts relate to each other and evolve.

In achieving cooperation, predictability is usually helpful, but this is not necessarily the case in conflicts: while playing tic-tac-toe, the player whose next move is predicted by his counterpart, usually loses the game. In general, for the purposes of defense it might be sufficient to build up a reputation of someone who can and is prepared to retaliate, although it may be left open what that retaliation would consist in. But if at least one of the parties is out for more there will be a strong incentive to attempt to surprise the other. Such surprises might bring victory, but might also lead into disaster. The problem with surprise attack is, that the enemy might only become manifest in retrospect, unless one could identify him by some hidden trait beforehand. It is no coincidence that the same difficulty resurfaces in Schmitt's answer to the question: Who is sovereign? Who determines the decisive friend-and-enemy grouping?

Expectedly, Schmitt's specification of sovereignty is not *stricto sensu* a definition: as usual, he prefers to argue from extreme possibilities (Schmitt 1996: 35). According to Schmitt, sovereign is the one who decides on the state of exception. Apparently this definition relies upon some sort of a *disposition*-predicate, though it seems to be a disposition that does not necessarily translate into behavior in a predictable way. It seems to be goal-orientated, depending on discretion rather than a transition governed by certain rules or a predictable effect of antecedent circumstances. Sovereignty therefore remains a secret whose revelation always comes as a surprise: even though constitutions might have already attached this label to a specific position, we will never know whether its holder is sufficiently inclined to live up to what we might vaguely identify as an adequate challenge. In 1936 Schmitt's contemporary and official enemy Rudolf Carnap, having managed in time to escape from Nazi Germany to the U.S., published an equally famous paper on dispositional predicates (Carnap 1953), that might shed some light on an asymmetrical bias inherent in Schmitt's technique of concept-formation. Abandoning the camp of logical positivism, Carnap has shown that dispositional predicates, like *beingsolvableinwater*, referring to something not yet realized and not immediately manifest, cannot be defined, but only grasped by a reduction sentence, that puts the disposition second: instead of asserting that *the sovereign is the one, who decides on the state of exception*, we can only state that *the person who decides on the state of exception is sovereign*. Putting it this way, the sovereign cannot be known in advance. He proves himself only in retrospect. And the reduction sentence only yields a conditional definition, i.e. no definition at all.[8]

There might be something else fishy with Schmitt's determination of sovereignty that has been anticipated by Walter Benjamin and later pointed out by Franco Moretti referring back to Benjamin (Moretti 1983: 46). Sovereigns that declare the state of exception—which almost by definition demands an asymmetrical counter-concept—more often than not turn into tyrants. Instead of considering his dictatorship to be of only limited duration and for a good cause, namely leading the country away from civil war, such a sovereigns not only occasionally lead the country straight *into* civil war and disaster by staying in power for too long and getting corrupted by it, confirming the fears of Jacob Burkhardt, Lord Acton and many others. Whoever is able to declare a state of exception will be exposed to the charms of power, and if this goes unchecked—as the notion of sovereignty implies—corruption *will* take place. Shakespeare's tragedies illustrate this point well by offering us a dynamic of events often diametrically opposed to that envisioned by Schmitt for the cases of sovereign dictatorships. Dictatorship does not eventually end the state of exception; it is much more likely to drift into civil war. Schmitt's sovereign does not save us from

8 This does not necessarily invalidate Schmitt's point, because (relational) disposition-predicates might be thought of to be much more common than they appear to be, and may indeed be the only predicates we have. Elaborating on this issue here would, however, lead us too fare astray from our volume's topic.

catastrophe, it brings forward disaster. All this, of course, is not Schmitt's fault, but the aura he gives to a tyrant seems misleading.

But the dispositional determination of sovereignty entails also a kind of paradox, putting a double bind on sovereignty. If the would-be sovereign could be driven into a situation where he has to decide on the state of exception, than he has already lost control and is not a sovereign anymore. Could the assumed disposition be tested and turn out to be positive, it would evaporate. There would be no sovereigns at all, we were to conclude, or there would be very many, namely all those who define contexts and frame situations, for example, lawyers working behind the scene. Despite being well aware of all this, Schmitt's sovereign is still in the end the hero who is called on stage when things go astray, that is, when he has already lost his grip of the situation and is no more a sovereign. This makes Schmitt's way of determining the concept look a bit bizarre. Taking this into account, we might call sovereigns all those who easily get irritated, may in response to this irritation cause many others real troubles, and who live by this reputation of a large-scale troublemaker. There might be many such persons, but one of them could also be a local primus inter pares; the threats and virtual negations by which these sovereigns hold each other in check are still not very well understood. Schmitt's notion of sovereignty was not meant as a tool to deal with these complexities; rather, it points at a certain fetish and surrounds it with a mystical or epiphanic aura in order to give it some orientation, against all odds.

A third strategy for incorporating a bias in a pair of counter-concepts is to define one member of the pair in *privatve* terms, describing it, so to speak, by its deficits.[9] After having defined men as living beings that have reason, Aristotle in the first chapter of his *Poli-*

9 Such or at least some privative predications might also be subsumed under dispositional ones. Holes might here serve us as an example. A hole is where something else is not. It needs a surrounding and is defined in relation to this surrounding. It needs a host, we might say. But it can also receive guests. You can put things into the hole, fill it, hide in it, leave it again or empty it. Holes, consisting of nothing, can perhaps only be specified by what you can do with them and what they allow for, i.e. by their dispositions. We should also not forget, that these dispositions should be conceived of as being ungrounded, as holes do not have parts that could be identified as the grounds for their dispositions (Casati/Varzi 1994: 110).

tics describes the social position of women, slaves, and children as correlates to their respective lack of reason. Today, we can only agree with him with respect to children, accepting them as human beings in spite of the fact that they still lack something essential. Another famous privative definition from our western tradition is the legacy of the Church Fathers who struggled against the Manichaean belief in two gods—the evil creator and the Redeemer who would save believers from the first one. In response, St. Augustine argued that evil is nothing positive but only a lack of being, so there is no additional creator standing behind it. Consequently, sin is nothing inherently bad but is merely a lack of goodness. Lacking goodness, men are dangerous and in need of a superior keeping them in charge. Schmitt follows suit. The Maoist Joachim Schickel once suggested to Schmitt that one could perhaps also try to conceive of the enemy as a special kind of friend (Schickel 1993: 9). Schickel proposed a dialectical maneuver, but we can also accomplish something close to it by a privative rearrangement. Friendship may be taken as the more encompassing term and the enemy could be defined by its privative specification: the enemy is a friend lacking something. He might know me all too well, but he lacks compassion and does not share my preferences. The optimist will put his energies into education or supplying other kinds of good hoping that the asymmetry will eventually disappear, whereas the pessimist will invest into the army and the police, hoping that this security beef-up would at least stabilize the status quo.

One specific counter-conceptual setup owing its asymmetry to an attribution of deficiency has been already commented on above: the concept of secularization, allowing us to distinguish between noble origins and illegitimate descent, or between original and copy. For Schmitt, the legal formulas of the omnipotence of the modern state are only *superficial secularizations of theological formulas of the omnipotence of God* (Schmitt 1996: 42). The metaphor of secularization portrays the history of ideas as if they were a kind of property that can be owned legitimately or used illegally, inherited in a lawful way or just be stolen. This presupposes a deficit, namely, the lack of legitimate property rights. It marks the non-irreligious counterpart in privative terms. But do we inherit what has been thought and written by past authors? Are we perhaps indebted to them? Can we appropriate their distinctions? The typically short polemical invectives that the concept of secularization invites, insinuate that the moralists, positivists, or legalists in question have stolen something and have thereby put on themselves a burden that they won't be able to carry.

Our fourth entry on this list of the possible inroads of asymmetry into counter-conceptual pairs entails questions of consistency. Two such questions we like to address. The first is: Are there consistency requirements in marking our enemies, and if not, will this lack of consistency invite asymmetry? Marking the enemy takes precedence in Schmitt's concept of the political. But how will you distribute your markings on a global map? Should you map the political actors thus localized in a consistent way? And how many different colors will you need? As we have already pointed out, being deemed consistent may be a disadvantage in the face of the enemy. Questions of coalition building add a further dimension to this. Someone guided by the maxim *my enemy's friends are my enemies* would perhaps be someone who aims at maximum *consistency*, but he would most likely also be a poor kind of enemy. Social pragmatics dictates that one should not try to be friends with all the enemies of one's enemy, and one should try to be friends with at least some of one's enemy's friends or friends of these friends (Martin 2009: 45). Such coalition-building machinations, however, make a political actor less predictable even to his friends, either turning him in the long run into a mild paranoiac or realistic loner, seeing all others as plotting against him, or forcing him to live with a certain amount of internal dissonance, while still having friends, or at least alter egos that you can call friends in public, though they might already have other plans.

Will you admit such inconsistencies to yourself, to your friends, to your counselors, to your national assembly, to the world at large? And how should you handle self-elevating attributions? Since Hegel's time social theorists are asked to apply to themselves the distinctions they prefer to try out on others. Will this make their theories collapse or force them to argue in specific ways? If the distinction between myth and reason is a reasonable distinction, than you have both myth and reason; if it is taken to be a mythical distinction, everything collapses into myth. The autological application of distinctions might be of some consequence and this brings us to our second question concerning consistency. Keen on drawing a line between friend and enemy, Schmitt insists that this distinction should fit the concrete situation instead of being taken as a ghost-like abstraction (Schmitt 1996: 30). Jacques Derrida wondered if this distinction between the abstract and the concrete is itself an abstract distinction or a concrete distinction (Derrida 2005: 116). Maybe it is a ghost-like distinction and Schmitt is haunted by the specter of the concrete that always allows you to speak from a privileged point of view, claiming that your opponents follow "ghost-like abstractions". To enjoy this self-privileging asymmetry, however, you should abstain from applying this distinction to itself, thereby failing the Hegel test.

VI Three More Varieties of Staging a Conceptual Asymmetry

The four entries in the above list of possible ways to arrive at a conceptual asymmetry derived from Schmitt and his critics, might look like a mere sophistry in comparison with the more robust asymmetries, we have briefly touched upon, namely those articulated by indexical distinctions, those arising from conflict over scarce goods that both parties cannot enjoy at the same time, and those that are due to the pragmatics of marking, that invites us to mark what is less common, while what is common goes unmarked. We have pointed at two criteria for identifying asymmetrical concepts. First, there must be a status difference or a situation of conflict; second, the actual relation captured by the relevant pair of concepts must lack mutual ratification by those participating in the relation. Non-ratified but socially relevant status differentials and social conflicts—in case they persist throughout many encounters—engender asymmetric concepts.

Instead of a conclusion, we would like to end this paper by pointing at three further variants of staging asymmetrical counter-concepts which seem to be of some relevance and which have not been integrated into the canon of asymmetrical counter-concepts till now. The first case is concerned with again with consistency requirements that seem to demand, that you either need to lie or else break off certain contacts. Should you, however, embrace neither of these two options, than you will be forced to address those, that you might like to convince by a double bind.. Jon Elster relates that Rosa Luxemburg held the "manipulative idea, that the working class can advance only by fighting for goals that the leaders have deliberately chosen because they are just out of reach" (Elster 1978: 59). Luxemburg assumes that it is only by defeats in minor struggles that workers would become more class-conscious and more radical. Should the socialist agitator then propagate goals that are just out of reach only to make the workers more class-conscious? Should one's answer be positive, this will certainly require hiding this maneuver from the workers. Should they, however, already agree and embrace the idea, further agitation would not be necessary. Jacques Rancière elaborated on a similar problem (Rancière 1999: 22-27). Rancière wondered, if Menenius Agrippa to whom Livy attributed the famous fable about the revolt of the body's members, could use this fable to calm down the revolting plebeians. If from the patricians' point of view the relation between them and the plebeians is based on the assumption that the latter need to be governed because they have no logos and therefore cannot govern themselves, than, of course, it would amount to a performative contradiction to try to convince them of their own lack of reason. If they could be convinced, however, the argument would collapse and there would have been no need to convince them in the first place. Rules of domination, legitimized among those in the superior position by a privatively defined counter-concept, cannot be legitimately ratified by the inferior party by appealing to the very same concept. To reduce dissonance or abstain from lying and self-contradiction, those who cultivate such conceptual asymmetries need to keep their contact with those assigned to the inferior position to a minimum, confronting them and their alleged lack of reason with a double bind. This *double bind* consists in the attempt to make them understand, that they won't understand, what is at stake.

Our second case is adapted from Freud's *Totem and Taboo*, where the story in question is further attributed to Darwin. According to Freud, in the old days humans lived in small groups tyrannized by patriarchs who had all the ladies for themselves. Naturally, the younger men did not like this; however, they did not simply hate the patriarch, but also envied him, which made it particularly difficult for them to revolt. Nevertheless, one day the youngsters in one of the groups rioted and killed the patriarch. But the new social order was difficult to establish as every one of them wanted to be the new patriarch. Eventually they established completely different rules. Accepting each other as equals, each man was allowed to have only one wife. This monogamy was supplemented by exogamy, establishing a system of international relations based on the exchange and abduction of women. Internally each community was stabilized by tabooing their patricide, which transformed the formerly competing bunch of brothers into a singular brotherhood. Eventually, the taboo absorbed the brothers' ambivalent feelings towards their father; furthermore, it transformed the murderous deed into a somewhat obscure sacrificial event, which was regularly commemorated by the feast reminding the brothers to keep the taboo.

Carl Schmitt tells a somewhat similar story in his reading of *Hamlet* (Schmitt 2009). Here the general problem is assumed to be the necessity to quickly identify the heir to the throne as soon as the death of the king or queen seems imminent, because a delayed enthronement may result in civil war. The successor may gain the power by dubious means, but, once in power, he or she must be obeyed, if civil war is to be prevented. One should not therefore look too closely at the details of the succession. This is what according to Schmitt, the play *Hamlet* and in particular the play in the play teaches its audiences: it makes the question of the queen's possible guilt in the murder of her former husband a taboo. To appreciate the play correctly, we need to understand its historical context, and the same, of course, holds true for Schmitt's own attempt at *hamletizing* another possible avenger. Already a few years before this intervention into the post-war debate on Germany's past, Schmitt argued that amnesty and amnesia, i.e. tabooing certain events of the past, will help to consolidate social order after war.

Building on Janet Beavin Bavelas et al. book on *Equivocal Communication* (Bavelas 1990) we can perhaps identify a general pattern here. Whenever someone confronts you with a set of alternatives which make you feel uneasy and which you do not want to choose from, you can indicate you unwillingness to be approached that way by an equivocal reply. The other side may now insist on your making a choice, risking conflict, or can tacitly withdraw, thereby allowing you both to pretend that the confrontation never occurred in the first place. This establishes a *taboo*. Such situations and confrontations with highly ambivalent alternatives might be more common and ubiquitous, than they appear to be (Giesen 2010). Tabooing establishes an asymmetry by reminding those who honor the taboo that they should not know what is tabooed, i.e. by a double bind. The *hamletization of the avenger* follows from the ambivalence and the ensuing taboo (see also: Derrida 2005: 169) and perhaps we should take this phrase to serve us as a metonymic label for the communicative constellation that is thus established. The taboo acquires its profile from what it hides from public discourse. The tabooed events might be talked about in private and they might even constitute a public secret, known to everyone, but perhaps not, at least not collectively known to be known (Taussig 1999). According to Schmitt, it is the single privilege of art to reflect on this asymmetry (see especially: Türk 2008: 87).

Our last case will bring us back to one of the examples we started from. In his analysis of land and lordship in medieval Europe Otto Brunner extensively dealt, among other things, with the feud. At the time when the modern state, with its monopoly on exercising legitimate violence, did not exist and the distinctions between private and public, army and police, foreign and national politics had no meaning, the feud, that we tend to look down upon as the law of the fist or a private warfare, was one of the more important institutions to achieve justice. Feuds, pursued by mounted knights, were aimed at retribution and vengeance to restore the broken order of what was considered to be right. In a similar vein, social anthropologist have argued that feuds can be seen as mechanisms to establish peace by allowing for social control, which would both serve as a deterrent against violent acts by making potential enemies fear violent retaliation, and offer a strong incentive to dissuade one's friends from engaging in violent acts for the same reason. Feuding parties met each other on an equal footing, engaging in negative reciprocity. Social order did not presuppose a state, which to Brunner and Schmitt already began to look like a thing of the past. Otto Brunner also extensively looked at a second institution: the household and its lord, which could be seen as the institutional nucleus of medieval society, characterized by relations that repeat themselves on larger scales: what the lord is to his household, the seigneur is to his peasants, and the prince is to all others standing under his protection. Brunner characterized this relation as one of exchange: the lord offered protection, and his subjects were loyal in return. Only a few years ago Gadi Algazi (1996) had put his finger on a possible link between these two institutions that Brunner did not consider with sufficient care, arguing that it was the feuding among lords that first of all made peasants seek protection. The lord offered protection against his competitors. This looks perhaps better than engaging directly in plunder and robbery or succumbing to a protection racket, but it also makes Algazi wonder whether one should describe this relation in terms of exchange and consent, as Brunner did. Speaking of protection and consent in this context casts the relation in the very euphemistic terms that also served as a cover for the early expansion of Nazi Germany to the east.

One might call this rhetorical maneuver *euphemistic altercasting*. It establishes an asymmetry in a somewhat paradoxical manner, namely by elevating those it discriminates against into a position of a greater responsibility. Their fate then becomes their own fault. Remarkably, this strategy of concept formation is also used by liberals. In Chapter V of his *Second Treaties* John Locke argued for a natural right of property that precedes the formation of government and is prior to any social conventions: things that we change by working on them and mixing labor with become in this way our property. This might sound just and plausible. However, David Hume, and before him Thomas Hobbes, opted for different explanations, making property a social convention, or something that followed from the social contract. But in America it was precisely Locke's notion of property that allowed for dispossession and violent eviction of native people who were apparently unable to make responsible use of the land and its resources. In contrast, the settlers coming from Britain did mix their labor or their servants' labor with it and began to fence it so that it could no longer serve as a public hunting ground (Damler 2008). According to Locke, each man is essentially responsible for his own economic fate. Some simply do not work hard enough, or so it seems. The haves, however, must not be very much concerned about this—it is not their fault. Here again, we see a maneuver of euphemistic altercasting at work. A social contract-thinker would perhaps attribute differences in property holdings to power differentials, whereas a liberal in the tradition of Locke will charge the have-nots with much more responsibility for their own fate and explain the asymmetry in a way they may not be very much inclined to ratify.

REFERENCES

Algazi, G. 1996. *Herrengewalt und die Gewalt der Herren im späten Mittelalter*. Frankfurt am Main: Campus.

Austin, J. 1970. *A Plea for Excuses*. In *Philosophical Papers*. 2nd Edition, by J. Austin. Oxford: Clarendon Press, 175-204.

Barth, F. 1990. *Cosmologies in the Making. A Generative Approach to Cultural Variation in Inner New Guinea*. Cambridge: Cambridge University Press.

Bavelas J. B. et al. 1990. *Equivocal Communication*. Newbury Park: Sage.

Blom, H. (ed.) 2001. *Quentin Skinner, Yves Charles Zarka, Hobbes—The Amsterdam Debate, Hildesheim*. Zürich, New York: Olms.

Blumenberg, H. 1983 *The Legitimacy of the Modern Age*. Cambridge, Mass.: MIT Press.

Blumenberg, H. 1985. *Work on Myth*. Cambridge, Mass.: MIT Press.

Brunner, O. 1992. *Land and Lordship. Structures of Governance in Medieval Austria*. Philadelphia: University of Pennsylvania Press.

Carnap, R. 1953. Testability and Meaning. In *Readings in the Philosophy of Science*, edited by H. Feigl/M. Brodbeck, M. New York: Appleton Century Crofts, 47-92.

Casati R./Varzi, A. C. 1994. *Holes and other Superficialities*. Cambridge, Mass.: MIT 1994.

Collingwood, R. G. 1939. *An Autobiography*. London: Oxford University Press.

Damler, D. 2008. *Wildes Recht. Zur Pathogenese des Effektivitätsprinzips in der neuzeitlichen Eigentumslehre*. Berlin: Duncker & Humblot.

Derrida, J. 2005. *The Politics of Friendship*. London: Verso.

Elster, J. 1978. *Logic and Society. Contradictions and Possible Worlds*. Chichester: John Wiley & Sons.

Feigl, H./Brodbeck, M. (editors). 1953. *Readings in the Philosophy of Science*. New York: Appleton Century Crofts.

Giesen, B. 2010. *Zwischenlagen. Das Außerordentliche als Grund der sozialen Wirklichkeit*. Weilerswist: Velbrück.

Goffman, E. 1981. Footing. In *Forms of Talk*, by E. Goffman, Oxford: Blackwell, 124-159.

Goody, J. 2010. *Myth, Ritual, and the Oral*. Cambridge: Cambridge University Press.

Gould, R. V. 2003. *Collisions of Will. How Ambiguity about social Rank Breeds Conflict*. Chicago: University of Chicago Press.

Grant, E. 1981. *Much Ado about Nothing. Theories of Space and Vacuum from the Middle Ages to the Scientific Revolution*. Cambridge: Cambridge University Press.

Hewitt, J. P. 1997. *Self and Society. A Symbolic Interactionist Social Psychology*. 7th Edition. Boston: Allyn and Bacon.

Hobart, M. 2000. *After Culture. Anthropology as Radical Metaphysical Critique*. Yogyakarta: Duta Wacana University Press 2000.

Holmes, S. 1993.*The Anatomy of Antiliberalism*. Cambridge, Mass.: Harvard U.P.

Horn, L. R. 1989. *A Natural History of Negation*. Chicago: University of Chicago Press 1989

Kelley H. H./Thibaut, J. W. 1978. *Interpersonal Relations. A Theory of Interdependence*. New York: John Wiley & Sons.

Reinhart Koselleck, R. 1985. The Historical-Political Semantics of Asymmetric Counterconcepts. In *Futures Past. On the Semantics of Historical Time*, by R. Koselleck. Cambridge, Mass.: MIT Press, 159-197.

Kuper, A. 2000. *Culture: The Anthropologists' Account*. Cambridge, Mass.: Harvard University Press.

Luhmann, N. 1995. *Social Systems*. Stanford: Stanford University Press.

Martin, J. L. 2009. *Social Structures*. Princeton: Princeton University Press.

Moerman, M. 1974. Accomplishing Ethnicity. In: *Ethnomethodology*, edited by R. Turner. Harmondsworth: Penguin, 54-68.

Momigliano, A. 1977. A Piedmontese View of the History of Ideas. In *Essays in Ancient and Modern Historiography, by A.* Momigliano. Middletown, Connecticut: Wesleyan University Press, 1-7.

Moretti, F. 1983. *Signs Taken for Wonder*. London: New Left Books.

Namier, L. B. 1961. *England in the Age of the American Revolution*. London: Macmillan.

Ogden, C. K. 1967. *Opposition. A Linguistic and Psychological Analysis*. Bloomington: Indiana University Press.

Piaget, J. 1971. *Structuralism*. New York: Basic Books.

Posner, E. 2000. *Law and Social Norms*. Cambridge, Mass.: Harvard University Press.

Rancière, J. 1999. *Disagreement. Politics and Philosophy*. Minneapolis: University of Minnesota Press.

Sahlins, M. 1972. The Sociology of Primitive Exchange. In *Stone Age Economics, by M. Sahlins*. New York: Aldine de Gruyter, 185-275.

Samuelson, P.A. 1966. A Pure Theory of Public Expenditure. In *The Collected Scientific Papers of Paul A. Samuelson. Volume II*, edited by J. E. Stieglitz. Cambridge, Mass.: MIT Press, 1223-1225.

Schickel, J. 1993. *Gespräche mit Carl Schmitt*. Berlin: Merve Verlag.

Schmitt, C. 1996. *The Concept of the Political*. Chicago: University of Chicago Press.

Schmitt, C. 2009. *Hamlet or Hecuba. The Intrusion of Time into the Play*. New York: Telos Press.

Smith, P. 2005. *Why War?* Chicago: University of Chicago Press.

Taussig, M. 1999. *Defacement. Public Secrecy and the Labor of the Negative*. Stanford: Stanford University Press.

Türk, J. 2008. The Intrusion. Carl Schmitt's Non-Mimetic Logic of Art. *Telos* 142: 73-89.

Linguistic Semantics and Historical Semantics

M. LYNNE MURPHY AND ROBERTA PIAZZA

INTRODUCTION

Reinhart Koselleck's engagement with notions from linguistics raises the question of whether there is a role for contemporary linguistic work in the practice of conceptual history—in spite of the fact that Koselleck's work "disregards the practice of almost all present-day linguists, semanticists, and analysts of discourses and ideologies" (Richter 1995: 45). We have approached this project as outsiders and have set ourselves the task of searching for the common ground between what we do as linguists and the interests of conceptual historians, and, in the process, raising critical questions where the commonalities are less than clear.[1] In this chapter we consider the notion of historical concepts with respect to the two levels of linguistic analysis at which the notion of a CONCEPT might be considered: lexis and discourse.

Much has changed in mainstream linguistics since the heyday of the European structuralist linguistics that informed Koselleck and colleagues' work—and much has changed in terms of what is considered "mainstream linguistics" since Richter's 1995 observations on the mat-

1 Our reading of Koselleck is, unfortunately, restricted to what can be seen through the lenses of translation (into English) and interpretation by other scholars.

ter. The notions of *langue* and *parole*, synchrony and diachrony, paradigmatic and syntagmatic have been challenged in various ways. The notion of *langue* as unitary and homogeneous has been repeatedly doubted and considered an unnecessary abstraction; similarly, the social aspect of language has been emphasized by challenging that *parole* reflects individual speakers' contributions. Current work on lexical opposition (e.g. Murphy 2006) questions the distinction between syntagmatic and paradigmatic, and work in Cognitive Linguistics sometimes flouts the synchronic/diachronic distinction (e.g. Sweetser 1990). Finally, as Fairclough (1989) points out, Saussure overlooked the crucial point that people have unequal access to and control of language. Since Saussure, the focus of linguists' attention has shifted from language as a system in its own right to language as a window on cognition and as the expression of social practice and a medium for social institutions.

Our starting point for considering the relevance of linguistic studies to "the semantics of historical time" (Koselleck 2004) is the vocabulary of social group identities, that is, socionyms (e.g. *black, white, gay, European*), the categories they represent, and the issues they raise. Socionyms plainly qualify as 'social and political concepts' in the Koselleckian sense in that they are socially and politically salient, consistently ambiguous, diachronically unstable and are constrained through their asymmetrical relations with counter-concepts, as Koselleck had in mind when he considered socionymic contrasts such as *Barbarian/Hellene*, and *Christian/Heathen*. From the lexical-semantic perspective, the study of socionym meanings is informed by general principles of categorization and, potentially, special principles of social categorization. From a more social-contextual approach, the attention is on the role that language plays in reflecting and constructing public beliefs and social ideologies, while the attention to contextualized textuality aims to reveal the semantic constructions that reflect the interest of particular groups in society. Both approaches start from a commitment to empirical evidence, whether they use data analysis to test hypotheses or they are instead totally driven by linguistic data. In recent decades, increased access to large, computerized corpora of linguistic data have aided some of the work we report below, and has led to challenges to formerly received notions about structure and meaning.

MEANINGS, CONCEPTS, AND RELATIONS AMONG THEM

We start by contrasting our lexical semantic approach to the semantic basis of Koselleck's approach—that is, the structural semantics that dominated European semantics in the mid-20th century. *Structural semantics* is itself a polysemous term (Coseriu/Geckler 1981), but at its root is Saussure's notion that "the sign by itself would have no signification" (Saussure 1959: 130). This leads to the "associational" structural semantics of his followers, such as Charles Bally (1940) who treated word meaning as an emergent property of a relatively unconstrained set of associations between words. This brand of structural semantics did not make it far in the linguistic discipline, since it lacks explanatory value. An unconstrained notion of relation is not valuable in trying to explain what can or can not be a word meaning. More interesting for us is what Coseriu and Geckler (1981) termed the "analytical sense" of *structural semantics*. This sense concerns the organization of vocabulary on the basis of oppositional relations, which in turn leads to the componential analysis of word meanings. Thus it is based on the assumption that "[o]nly on the solid basis of [...] a paradigmatic semantics can a well-founded combinatorial semantics be built up" (Coseriu/Geckler 1981: 19). This gives rise to the semantic field approaches with which Koselleck was familiar (particularly Trier), in which constraints on the properties of semantic fields give rise to and thus constrain word meanings.

The primary contrast between this approach and the approaches that followed in the later 20th century to today is the former's focus on the language itself as the object of study—the thing that is structured. In contrast, many (if not most) contemporary lexical semanticists would see their job as discovering

> "how linguistic utterances are related to human cognition, where cognition is a human capacity that is to a considerable degree independent of language, interacting with the perceptual and action systems as well as language" (Jackendoff 2006: 355).

Linguistic semantics has, in large part, become the study of the structure of thought, that is, conceptual structure, and how that structure maps to language. As such, we should expect that the structure of any semantic field is constrained by the same organizational principles, but we might wonder whether any particular semantic fields have further constraints placed on them by the nature of the things that they denote. It is with this question in mind that we frame our discussion of socionyms.

But at this point, we must recognize the terminological distinction between psychological concepts and the Koselleckian concept. In defining his notion of CONCEPT, Koselleck (or at least, his translator) writes: "A word can become unambiguous. In contrast, a concept must remain ambiguous in order to be a concept" (Koselleck 2004: 85). This seems to imply that a difference exists between words and concepts and that only or especially the latter can be identified by linguistic-semantic criteria, on the grounds of unresolvable ambiguity. Yet on the linguistic view, most words are polysemous, their senses are fuzzy, and only under extraordinary circumstances can a precise sense for a word be resolved in context. Take, for example, a word like *date* as in *Mary has a date with John tomorrow*—it is likely that Mary, John, the speaker, and the audience for this sentence will have different ideas about what *date* means in this context—and it is not even necessary that any of the participants or referents in this exchange have a single discernable sense for *date* in mind when the sentence is uttered or reflected upon. Even concrete nouns like *book* or *coffeepot* can remain ambiguous in context as to whether they should be construed with reference to their contents, their form, or their functions. Presumably, though, these are the types of notions that Koselleck intended to dismiss as mere words, rather than concepts. On our view, rather than trying to distinguish socio-political conceptual terms from 'mere words' on the basis of linguistic-semantic properties, they must instead be distinguished on the basis of the non-linguistic forces that determine or constrain the extent and the 'directions' of the ambiguities.

The disconnect between linguistic semantics and conceptual-historical semantics is illustrated by recent work on the name/concept *Europe*. In the prominent linguistic-philosophical treatments of proper names in the 20th century, *Europe* should be among the least ambiguous of words, since, as a proper name, it can be held to be a "rigid designator" (Kripke 1980)—an indexical term that has no sense, just the potential to refer. But as illustrated by Geoffrey Williams, in his work on the representation of EUROPE in the French press (Williams 2007; Williams/Piazza/Giuliani, in press), even proper names like *Europe* can be polysemous and should be understood as embodying the prototype model of meaning (based in Wittgenstein's (1958) notion of family resemblance) so as to account for its variation of meaning in context. In terms of the media representation of Europe in the British, Italian, and French press this means that

> "whilst different senses [of *Europe*] can be isolated, they are all associated only through a continuum with different aspects being activated as contexts change. [... N]o single meaning can be accepted as generally prototypical, and it is necessary to accept that a variety of Europes may be under discussion at any one time" (Williams/Piazza/Giuliani, in press).

This view of *Europe* puts it squarely in the realm of the Koselleckian concept. But from a linguist's perspective, Williams' observations mean interpreting *Europe* as being *more* like other (non-proper) nouns, such as *date* or *coffeepot*. That is, linguistically speaking, *Europe* is not a special word because of its ambiguity—on the contrary, its ambiguity makes it typical (though no less interesting).

The assumptions that word meanings are fuzzy and context-driven of course has implications for the relations among words or concepts. Recent work on the relations between words (or their senses), such as lexical opposition, are context-dependent and ever-changing (e.g. Murphy 2003; Croft/Cruse 2004). Pairings of words in relations such as antonymy may become conventionalized through use—they become part of our language community's 'antonym canon'. But in order to succeed as a canonical pairing, any two words have to fulfill the criteria for a good opposition in context. This criterion (following Murphy 2003) is minimal, relevant difference (and maximal similarity) across a range of salient contexts.

So, for example, *heterosexual* is the canonical antonym for *homosexual* because:

a) in contrast to other candidate concepts/words (*bisexual*, *transsexual*, *asexual*, etc.), *heterosexual* conventionally denotes a sexual identity that is most like *homosexual*, in that like *homosexual* (but unlike the others) it denotes an identity predicated on single-sex object-orientation;

b) *heterosexual* is nevertheless incompatible with *homosexual*, in that the value of the object-orientation is 'other sex' rather than 'same sex';
c) the (a) similarities and difference (b) are what is most relevant to the context at hand (this abstract discussion of sexual identities), and
d) the particular relevance of difference (b) has been conventionalized, in part through repeated co-occurrence of the pairs in contrast.

On this view, *heterosexual* and *homosexual* have meanings in their own right; that is to say, the word meanings do not emerge solely from the fact that they are opposed in a semantic field. Another way to put this is that they have a 'positive' definition, not just a 'negative' one.[2] This position on word meanings and relations contrasts with the structuralist semantic notions that may have informed Koselleck's thinking. Structuralist semantic approaches to have been put into practice and tested as computational models of meaning, for example WordNet (Miller 1990; Fellbaum 1998) and Latent Semantic Analysis (Landauer/Dumais 1997) and have been found incapable of representing the fine points of meaning that human users create and rely upon (see Sampson 2000 and G. Murphy 2002, respectively). On our view, it is pragmatic principles, rather than semantic structures that determine which words are used as opposites in any context. At the same time, there is a social and textual 'super-context' that ensures that *heterosexual/homosexual* is the most salient opposition in most contexts and repeated exposure to these words as a pair has the effect of maintaining that salience across many contexts.

New oppositions are recognized through a number of discourse structures. Jones (2002, 2007), for instance, identifies a number of such structures for English such as *X, not Y*, *more X than Y*, and cases of grammatical parallelism in which a more familiar opposite set supports a novel one (*Kennedy* DEAD *is more interesting than Clinton* ALIVE; Jones 2002). These are argued, in Murphy (2006) and Jones et al. (forthcoming), to constitute grammatical constructions—that is,

2 *Positive* (versus *negative*) *definition* should not be confused with the notion of positive (versus negative) evaluation of a definition in a social context—an issue that is raised later in this chapter.

delexicalized yet meaningful linguistic structures. Such structures, which Jeffries (2010) calls "replacive" oppositions, have been shown to be present (if sometimes used to achieve slightly different effects) in Swedish (Murphy et al. 2009), Japanese (Muelheisen/Isono 2009), and from the earliest stages of grammatical language in children (Murphy/Jones 2008). Davies (2007) and Jeffries (2010) have recently used these observations to investigate the ways in which novel linguistic oppositions are used to effect particular ways of conceptualizing political situations.

ASYMMETRY AND CONTRAST

In "The historical-political semantics of asymmetric counterconcepts" (2004), Koselleck specifically defines his focus as being on the contrasts that bisect humanity into two mutually exclusive groups—that is, 'us' versus 'everyone else'. It is unclear from that work whether that restriction was introduced as a convenience for the purposes of exposition or a prerequisite of identity-concepthood, that is, a claim that only such grand divisions of humanity could count as concept and counterconcept. Such neat dichotomies, such as between *Barbarians/Hellenes* and *Christian/heathen*, display asymmetry *par excellence*, in that they can be defined in terms of a positive concept and its negation. That is, *Barbarian* and *heathen* cannot be defined without including 'not Hellenic' and 'not Christian', respectively. The negated concept can never be the positive concept's equal, since it is conceptually more complex—involving the positive concept plus negation—than the positive version.

The types of oppositions that inform our identities everyday are more complex in their relations to one another. While they may be based in the grand scheme of 'us' versus 'not-us', the primary *lexical* expressions of notions such as RACE or NATIONALITY are rarely dualistic. Most of the ways in which we identify ourselves are not 'us' versus 'everyone else', but 'us' versus 'those people', and the complexities of such relations heighten the relevance of the notion of asymmetry—since more limited divisions arise not through the logical necessity of asymmetrical 'us/not-us', but an 'us/them' whose asymmetry may not be reflected in different group sizes or even, necessarily, differences in power and social status. It certainly reveals something about a culture when the 'not-us' concept is lexicalized. At this point other people are conceptualized as being a group, rather than as just any collection of individuals who are 'not us'. But at the same time it is also interesting to investigate whether asymmetries in social/political categorization arise from the lexicalized terms that do not start out as logical asymmetries. If two terms can be positively defined (recall the *heterosexual/homosexual* opposition above), yet display linguistic and conceptual asymmetry, then this type of asymmetry requires further interrogation.

From here, we explore asymmetry with respect to the ways in which linguistic-semantic structures enforce or enable duality and asymmetry, the further creation and realization of asymmetries through textual/discourse means, and finally the cognitive underpinnings of asymmetries as they apply to socionyms.

SEMANTIC ASYMMETRY IN LINGUISTICS

Much of language structure (sounds, word forms, sentence structures) can be described in terms of binary contrasts. Modern linguistics has (not uncontroversially) made much of these dualities and the asymmetries within them through the concept of MARKED OPPOSITION, originally promoted in the Prague School of linguistics with reference to phonological and morphological analysis (e.g. Trubetzkoy 1939). Within a marked opposition, one member of the pair is essentially the more "basic" or "natural" member of a pair for a specific context. To take a morphological example, we have a contrast between singular and plural. In English, we have *cat* and *cats*. The singular form is the basic form of the word—it is UNMARKED with respect to the plural form, which is said to be MARKED. This is generally true across languages; if a language has different noun forms for singular and plural, it is usually the plural that deviates most from the basic form of the noun, and it is never the singular that is most deviant. However, markedness relations are always context specific; so for some contexts, for instance right after the word *many*, we can say that the *plural* is less marked than the *singular*. Besides differences in formal complexity, differences in frequency and breadth of distribution are the other symptoms of markedness asymmetry. Breadth of distribution can include instances where the unmarked member of the pair is NEUTRALIZED with respect to the contrast with the marked member. For example, we use the singular form in some cases when we're not referring to just a single member of a category, as in *The domestic cat is a distant relative of the lion*, which concerns more than just a single example of a cat or lion.

So far, this consideration of markedness lies entirely at the level of language description. The notion becomes much more theoretically powerful when it is used to *explain* these asymmetries. Here the assumption is that formal markedness reflects complexity or deviation at non-linguistic levels of consideration. In the case of singular and plural, the plural denotes something more complex (cat+cat...) than the singular does (cat). In other words, items that are more marked in terms of their distribution and linguistic complexity are hypothesized to be more cognitively complex.

Applying and representing markedness in the semantic realm has always been controversial, but at the same time it has served as a convenient shorthand in describing distributional differences among lexical items or grammatical forms that seem to reflect conceptual asymmetries. Talking about opposition between lexical items, such as *Christian/heathen*, involves also thinking about their relationship to opposition through negation, that is *Christian/not Christian*. A major division in thinking about the nature of negation is in whether one considers the relation between the negative and the positive to be an inherently asymmetrical relation—for the negative to be marked and the positive unmarked (Horn 2001). Michael Israel stands on the side of asymmetry:

> "In principle, opposed terms must be equal in their opposition: one term cannot be more opposite than another. But in natural language opposites are never equal. There is a consistent imbalance between the unmarked expression of affirmation and the marked expression of negation; between the general utility of affirmative sentences and the pragmatically loaded uses of negative sentences; between the simple logic of double negation and the not uncomplicated pragmatics which insures that denying a negative is never quite the same as asserting a positive" (Israel 2004: 701).

There is the question, however, of whether all forms of opposition consist of a 'positive' and a 'negative'. We have seen this already with *heterosexual* and *homosexual*. Each can be defined in a positive way, that is, without recourse to the negation of the other. So, while *homosexual* entails 'not heterosexual', 'not heterosexual' is not a sufficient definition for *homosexual.* Instead, it must be defined in a parallel way to heterosexual—'having a sexual orientation towards members of the {same/opposite} sex'. Greimas's (1983) semiotic square helps to illustrate this point, in that it "carefully distinguishes different *ways of setting up oppositions*" (Heins 2004: 512, applying the square to a historical-semantic analysis) and thus reveals the interplay between negation and antonymy as means of achieving opposition in language. As figure 1 shows, a term can be in a contrary relation with a lexicalized opposite (antonym) as well as a contradictory relation through negation.

Figure 1: Semiotic square

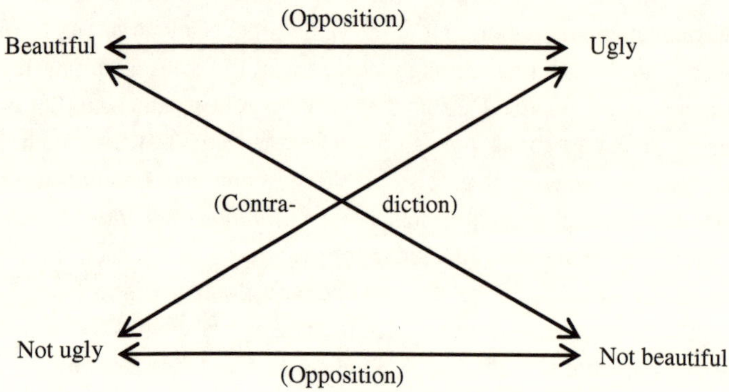

Source: Høstaker 2005

The complementarity between a positive expression (like *ugly* in figure 1) and the negation of its lexical opposite (*not beautiful*) is not equivalence, and indeed, psycholinguistic experiments show that people generally perceive negations of gradable adjectives (i.e. those that describe properties that can be had to greater or lesser degrees) like *not ugly* as meaning something different from their affirmative opposites, such as *beautiful* (e.g. Paradis/Willners 2006). We expect this to be the case when the two positive terms describe gradable properties. But, at the same time, gradable opposites are often reinterpreted as absolute, dichotomous categories, as is the case for *black.* When *black* is used to describe people of sub-Saharan African ancestry, is complementary to, but not synonymous with, *not-white*. But *black* is also sometimes used to *non-white*, as in:

"The number of black students increased to 5931, 32% of the total. Of these about two-thirds were African and most of the balance Indian" (University of the Witwatersrand, *Annual Report of the Vice Chancellor 1993*: 3; quoted in Murphy 1998).

In the case of typical contrary socionym oppositions, such as *black/white* and *heterosexual/homosexual*, there is less clearly a case of concept X and counter-concept not-X than there is for, say, *Christian* and *heathen* (= 'non-Christian')—the asymmetry is not semantically inevitable. Nevertheless, we still have a strong social sense that one of these is the "unmarked" concept, that is, the norm or default, and if there is a sense of the "exclusion of the middle" or of other possibilities, then it is the more socially "marked" category that can be interpreted as the negation of the unmarked. That is to say, that *black* can be interpreted to mean 'not white', as in the Witwatersrand example above, but *white* has not gained a sense 'not black' in the mainstream culture. Similarly, *queer* can be inclusive of both 'homosexual' and the "middle" category 'bisexual', but *straight* is not interpreted as having a sense that is equivalent to 'not exclusively homosexual'. This is a different kind of polysemy from that of unmarked terms, for which neutralization of contrast is sometimes possible. So, for example, the masculine *man* may be neutralized so that it refers to all of humanity, regardless of sex differences (*man has evolved as a tool-maker*). The distinction between *black* and *white*—and the status of *white* as denoting the unmarked case, on the other hand, is reinforced by the more general 'not white' use of *black*.

A signal of semantic markedness *sine qua non* is asymmetry in lexicalization—that is, one category is "marked" by having a word for it, and the other goes unmentioned. We see this repeatedly in the history of socionyms, with the lexicalization of a term for the "deviant" social group in the absence of a lexicalized term for the "norm". For instance, in English *black* as a racial term historically precedes *white* as a racial term (Flexner 1976) and the classification *heterosexual* was only invented two decades after *homosexual* in German and Hungarian[3] (Dynes 1985). The deviations from accepted norms are lexicalized, but they cannot be understood as deviations unless there is an underlying conceptualization of the "norm".

But this raises for us a problem in interpreting Koselleck's view of concepts, which holds that:

"Each concept is associated with a word, but not every word is a social and political concept" (Koselleck 2004: 84).

"From the concept of the one party *follows* the definition of the alien other." (Koselleck 2004: 156, emphasis added)

"Signifier and signified coincide in the concept insofar as the diversity of historical reality and historical experience enter a word such that they *can receive their meaning only in this one word, or can be grasped only in this one word.*" (Koselleck 2004: 85, emphasis added)

Here, of course, we run into a translation problem between the use of *concept* in linguistics and in Koselleck's work. For Koselleck concepts "are founded in politicosocial systems that are far more complex than would be indicated by treating them simply as linguistic communities organized around specific key concepts" (Koselleck 2004: 76). For us, on the contrary, a concept is any psychological representation that has the *potential* to be lexicalized, but not every concept is associated with a word (G. Murphy 2002). This understanding of CONCEPT includes the *ad-hoc* concepts (Barsalou 1983) that we invent in particular contexts and which may never be lexicalized, as well as well-entrenched concepts that are lexicalized. But in the case of *heterosexual* or *white*, the "unmarked" identificational concept predominant in the culture has only been lexicalized in reaction to the existence of a term for 'the other'. In order for 'the other' to be recognized as 'other', there must be some incipient concept of a delimited 'us'. The concept must exist on the social-political and cognitive plane well before it is a word. But on Koselleck's description, concepts as a rule are embodied by words. Richter (1995: 9), on the other hand, writes that "German conceptual [i.e. following Koselleck] historians distinguish concepts from words.

---

3 Both terms entered English simultaneously via translation of Krafft-Ebbing's *Psychopathia Sexualis.*

A concept may be designated by more than one word or term". It is difficult to reconcile this with the unambiguous statement of the final Koselleck quotation above, and it does not offer the possibility that a socio-politically important concept may be indicated by the absence of a word.

The disconnection between social history and linguistic/conceptual history, and the priority of the former over the latter is acknowledged in Koselleck's work. But since social history is mostly described as taking the form of events, we must assume that the unspoken, unmarked conceptualizations that inspire marked counter-concepts indicate a closer interrelation of event history and conceptual history than is afforded by equating the conceptual and the linguistic. While the keywords of a culture are a major means by which we can access a conceptual history, they are as much evidence of the conceptual unsaid as the said.[4]

This leads us back to the question of whether asymmetry is a necessary property of oppositions, or whether the asymmetries are created in the social realm rather than the conceptual. There is a linguistic tradition of thinking of antonym relations as inherently asymmetrical, so that Vendler (1963), Givón (1970), Ljung (1974), and Handke (1995) have defined antonymy as a relation between marked and unmarked lexical items, thus making asymmetry a co-requisite of opposition. Such grand claims about the necessity of asymmetry arise from a very narrow view of what an antonym is; in this tradition *antonym* is defined so as to include only gradable predicates (usually adjectives) in logically contrary relations, such as *tall/short*, *young/old*. In these examples, physics determines the asymmetry—measurements in these dimensions have a fixed starting point (e.g. the ground in the case of height, birth in the case of age), but no fixed end point.

But if we cast our net more broadly among conventionalized semantic opposites, the semantics of the opposed words do not *ab initio* necessitate asymmetry. Consider *north/south*. On a purely physical plane, directions on a globe in space are symmetrically related, yet we still find linguistic evidence of asymmetry—for example, the usual

4 Among linguistic semanticists, Anna Wierzbicka (e.g. 1997, 2006) in particular has pursued a keywords approach. While her theoretical model and metalanguage are not uncontroversial, her methods result in insightful observations of the data.

conjunction of *north and south* (439 occurrences in the British National Corpus) rather than *south and north* (11 occurrences). In specific contexts, such as *northerner* versus *southerner* in the UK, other markedness relations may obtain, with different types of linguistic evidence—for example, the perceived need to mention a person's northernness or southernness, demonstrating the contextual asymmetries.

Clearly, contextual forces are at work in many cases of oppositional asymmetry. From here, our task is to consider how such asymmetries arise and are maintained in language, from current discursive and cognitive perspectives.

DISCURSIVE SOURCES OF ASYMMETRY

What determines asymmetries: do they depend on the language itself or, rather, on who uses it (Thornborrow 2002: 8)? Modern discourse studies, mainly informed by Michel Foucault, view discourse not simply as a set of signs but as one of many aspects of social practice, or as "practices that systematically form the objects of which they speak" (Foucault 1972: 49). From this it follows that discourse is "something that produces something else (an utterance, a concept, an effect), rather than something which exists in and of itself and which can be analysed in isolation" (Mills 2004: 15). This reality-shaped and -shaping notion of discourse determines the political charge of modern discourse studies whose theorists and practitioners distance themselves from more traditional Marxism-informed notions of ideology and favour an emphasis on discourse that recognizes people's greater agency in the dominance process. Critical Discourse Analysis (Fairclough 1989, 1992; Wodak 1989, 2009; Krzyzanowski/Wodak 2008; van Dijk 2008; van Leeuwen 2008 among many others) as well as Positive Discourse Analysis (Martin 1999), with their debts to Habermas (1984), Althusser (1971), Bourdieu (1992) and Foucault (1977, 1980), have explored the role of language in perpetuating inequality, focusing on oppositions that powerful groups have transformed into asymmetries, whereby a term is subordinate to the other, and in which synonyms construing exclusion, relegation, marginalization and derogation proliferate. Studies like Poole's (2002) investigation of the representation of Muslims in the British press confirms that, while the image of Islam is not a unified one, Muslims are mostly portrayed as demonized "others". She found that only 15% of the articles about Muslims in her 1993-97 corpus were about British Muslims, which underscored the element of "foreignness" in more common stereotypes about this group, presenting it as inhabiting a very distant geographical space rather than the UK. Representation, therefore, is not a direct reproduction of objective reality; there is always "a mediating effect whereby an event is filtered through interpretive frameworks and acquires ideological significance" (Poole 2002: 23). Clearly, asymmetries that are reified through representation via language arise through socio-historical events and the distribution of and access to power in society. Still on a large scale, *Europe* discursively resulted from the "dichotomizing discourses of orientalism" in opposition to non-Europe (Gal/Irvine 1995). In a political domain, Fairclough's (1989) classic discussion of Margaret Thatcher's rhetoric in a 1985 interview with Michael Charlton highlights the alternation between inclusive *we* referring to the electorate and audience, and exclusive *we* indicating a collectivity, like the state. Together with *we*, *you* encourages solidarity amongst the people on the basis of their common experience. Thatcher often fudges the opposition between the government and the ordinary people and her reified *us* community is clearly exclusive of 'others'. More explicit linguistic construal of asymmetries refer to immigrants and refugees, always represented as antagonistic *them* to *us* (KhosraviNik 2010).

Fairclough (1989: 116) observes that "words are likely to have such relational values simultaneously with other values". For instance the use of racist vocabulary has "experiential value" in relation to the negative representation of a specific group; however "its use—and the failure to avoid it—may also have relational value, perhaps assuming that racist ideology is common ground for the speaker and other participants" (Fairclough 1989: 116). In any case, the creation of asymmetries in discourse operates at the level of "representations" and the production of meaning through language. Representations are approached in different ways. They can be viewed as mirrors of events or ideas in the world, or vice versa can be seen as especially produced by the individual people who use them. Alternatively, a "constructivist" or "constructionist" approach recognizes the social meaning of language and believes that meaning and representations are socially/collectively constructed and negotiated through a continually dynamic process (Hall 1997).

Asymmetries are most typically expressed in public or institutional discourse in which power relations are "exercised and enacted" through a number of "discourse types" that reflect social-cultural codes and conventions of the dominant groups (Fairclough 1989: 43). In a CDA tradition, for instance, van Leeuwen (2008) emphasizes the role of "assimilation" that is realized by plurality when referring to the media and in general public discourse around *Muslims* or *immigrants* and that provides the basis for asymmetrical oppositions against 'us'. In the text from the 12th May 1990 *Sydney Morning Herald* he analyses, immigrants are represented as assimilated and treated as statistics, as the subject of economic calculations but also as the cause of the Australians' fear:

"We have avoided most of the problems that bedevil Western Europe because few of our non-European migrants have been poor, black, unskilled, Muslim, or illegal" (van Leeuwen 2008:39).

The 'us' versus 'them'–'the others'–'the unwanted' opposition is the asymmetry at the basis of racist discourses (see the Duisburg group, committed to the identification/documentation/opposition of racism in Germany, and its media analysis of such right wing anti-foreigners newspapers as *Bildzeitung* and the conservative quality German press; Wodak/Reisigl 2003). Wodak (2004) illustrates her discourse-historical approach with an analysis of a 1992 Austrian petition proposed by Haider's party, which listed measures to "secure a fatherland for all Austrian citizens" and protect them from the incipient siege of immigrants. Wodak highlights the presence of "criminonyms" ('aliens') and "actionyms" ('immigrants') in the text of the petition, which, in conjunction with a reference to expected discrimination *topoi*, portrays the "others" as invasive in opposition to an emotional notion of *Heimat* as home of the "real" Austrians.

Asymmetries construed along the 'us' *versus* 'them' opposition are conveyed discursively and multimodally as well as for instance in the discourse of the media that echoes people's beliefs while at the same time reifies particular realities through language. Investigations by van Leeuwen (2008) and Lipson (2009), among others, point out how Allied soldiers in the Gulf war and in the 2003 Iraq war, respectively, were portrayed as individuals while Iraqi soldiers were represented as an often disorderly group (this is close to Piazza's 2009 linguistic study of representation of coalition versus Iraqi forces in BBC and CBS war reporting). Such visual choices suggest that Iraqi, the 'others', are the same indiscriminate bunch and any differences are irrelevant. "Visual texts, like verbal ones, are filmed and edited through ideological and cultural filters and often call on viewers' cultural background and knowledge for interpretation" (Lipson 2009: 166). A representation of the Iraqis as an uncontrollable group and the presence of other editing devices (e.g. the frequent lack of synchronization between verbal and visual texts, Lipson 2009) contribute to construing the visual 'us' *versus* 'them'–'the others' asymmetry.

Inequalities in a culture can thus be represented linguistically and through public channels these inequalities are further realized. These means for creating, maintaining and subverting asymmetries are available for any socionym, whether it refers to 'us' or not. So, for example, in the following example, we can identify an asymmetry between *Africa* and *the Philippines* in that one is a country, the other a continent of more than fifty countries which the writer has failed to identify.

"If we allow them to bring down the government, we will be like the *Philippines and Africa*." (http://bangkokpundit.blogspot.com/2008_08_01_archive.html, spelling corrected)

But the mechanics and motivations for asymmetry are more basic when one of the opposed words refers to oneself, because, of course, the dynamic of *us* versus *them* enters into play.

Asymmetries, generally lexically or syntactically indexed, can also be expressed by oppositional meanings determined by pragmatic implicatures, for example irony. The general view is that irony synthesizes the contrast between appearance and reality in that it corresponds to the opposite of what one means. A very pertinent example of irony, Lady Macbeth's question, *Are you a man?* "implicates the impolite belief that [Macbeth] is so lacking in those characteristics which she perceives as male that his gender is called into question" (Culpeper 1996: 365). The early view that an ironic utterance means the opposite of what appears on the literal level (Searle 1991; Brown/Levinson 1987; among others) has long been challenged and Kaufer (1981) points out that irony contradicts contextual expectations not solely the literal form of an utterance. An alternative view (Sperber/Wilson 1981;

Sperber 1984; among others) sees irony as negating a prior original utterance and being an "echoic mention" or "echoic repetition" of it. An ironic utterance as "These people really know what they are talking about" uttered by a speaker who intends to cast a doubt on someone's knowledge or professionalism expresses the person's critical relationship with what someone else has said even in the absence of clear linguistic indicators that is even if these oppositions are not linguistically but inferentially realized. Thus the fictional masculinity in film (Piazza 2010) can be viewed as intrinsically located in interaction and discourse. In the American and Italian western genre, the good cowboys define themselves in antagonism/opposition to the negative gunslingers by means of an ironical discourse that critically echoes the exact words/utterances, the underlying propositions or the frames that are associated with the bad men.

In conclusion, discourse that is produced and appropriated by an empowered or disempowered group reinforces social oppositions and crystalizes them into an 'us/them' dichotomy in which one party is less valued than the other as for the Koselleckian *Barbarian/Hellene*, and *Christian/heathen* pairs. Asymmetry enters issues of identity and the representation of self and other both in real life and in fictional discourse when pragmatic strategies function to contrast a positive self against a negative other. The asymmetrical disparity between powerful and powerless groups, which discourse as a social practice expresses, however has no ontological permanency, just as the notion of 'markedness' above can only be made with reference to context, therefore such politically committed approaches as DCA and PDA aim to identify how language perpetuates alienation and marginalization and share with *Begriffsgeschichte* the potential for "identifying and combating ideologies" (Richter 1995: 53).

COGNITIVE SOURCES OF ASYMMETRY

Koselleck's diagnostic criteria for concepthood (ambiguity, opposition, and some aspects of asymmetry) are not special properties of socio-political concepts, but rather part of the nature of using language meaningfully. So our final question is to what extent the 'specialness' of asymmetrical counter-concepts can be upheld—that is, are they any different from any other type of meaning? Can we discover the roots of the asymmetrical power dynamics, discussed in the discourse-analytic tradition, in order to predict the asymmetry of categorization—which contribute to asymmetry of power?

Linguistic markedness is often held to mirror conceptual 'markedness'. In this case, formal elaboration in the language is taken as a signal of conceptual complexity (e.g. Givón 1991[5]) or negative affect (e.g. Yariv-Laor/Sovran 1998). More recently, Haspelmath (2006) has argued that markedness effects are the product of frequency and (neo-) Gricean communicative principles (such as those of Horn 1984 and Levinson 2000), which predict that communicators will act to make their linguistic contributions appropriately relevant and informative for their contexts. We are led to wonder if these positions can be reconciled to some degree for socionyms, in that the most frequent/familiar experience is oneself, and oneself might be argued to be cognitively simpler and more available than others—that is, there is a me and a not-me. At the same time, there is within an ingroup a communicative need to label the other, but not the self or the ingroup, because the ingroup norm can be taken as background information. So, if, a Hellene said *I need to find someone to marry*, the idea that the 'someone' is a fellow Hellene is implicated. To say *I need to find a Hellene to marry* would be over-informative in a normal context, since the ingroup default can be assumed. But if a barbarian was who one wanted to marry, *someone* would not be informative enough—the barbarian must be linguistically realized.

The ambiguous and asymmetrical nature of 'us/them' identifications is well accounted for on a cognitive level through Social Identity Theory (Tajfel/Turner 1979; Tajfel 1981) and its offshoot Self-Categorization Theory (Turner 1984), collectively known as *the Social Identity approach* (SIA). These social-psychological theories attempt to account for group behaviour "in terms of concepts that articulate societal and psychological processes that recognize the primacy of society over individual" (Hogg/Terry/White 1995: 262). Here, social identity is defined as: "that part of an individual's self concept which derives from his knowledge of his membership of a group (or groups) together with the value and emotional significance attached to the

5 "The marked category tends to be cognitively more complex—in terms of attention, mental effort or processing time—than the unmarked one." (Givón 1991: 337)

membership" (Tajfel 1981: 255). The approach holds that we maximize differences between groups and similarities within groups (following from general principles of categorization, which allow us to put meaning to our experiences). It also holds that humans are driven by a basic need for a positive self-concept, and therefore they have a favourable bias toward the group(s) to which they belong—that is, the groups from which they derive a significant part of their identity. This enhancement of the ingroup is guided by subjective belief structures (including ideological constructs) about the relationship between one's own and other groups. Social identity also entails de-personalization, that is, individual behaviour and perceptions of other individuals are made consistent with perceptions of groups to which the individual is assigned.

SIA is compatible with a social-constructivist view which denies the existence of "an absolute self, lurking behind discourse" (Benwell/Stokoe 2006: 3–4). Identity therefore is continually negotiated and subject to the changes imposed by the public domain and cognitive necessity. However, "[t]he contemporary social identity approach rarely has much to say explicitly about communication issues and is ripe for development in this direction" (Hogg/Reid 2006: 9). Murphy (1997) took a first step toward marrying SIA and linguistic semantics, positing and testing five hypotheses concerning social categories and semantic change, based on socio-cognitive hypotheses in SIA, and testing them in the semantic field of sexual orientation socionyms (with some comparison to race socionyms). The following are slight revisions to the hypotheses in that work:

1. If a socionym is adopted as a marker of identity, a lexical opposite will be introduced.

This hypothesis is based on the premise that the existence of a social category depends on the existence of another category to compare it to (Tajfel/Turner 1979), and Murphy (1997) found this to be the case and "[f]urthermore, the contrasts tend to be dichotomized, giving every label a unique antonym" (Murphy 1997: 43). Hence, once *bisexual* became a socionym around which an ingroup organized itself (and not just a label for a type of behaviour), it was no longer sufficient from an ingroup perspective to contrast *bisexual* to both *homosexual* and *heterosexual*, and so it was not long before *monosexual* was coined (in many places, nearly simultaneously) to fill the lexical gap that bisexual identity necessitated.

2. In the absence of broad-based changes to belief structures and socio-political status, new category names that are incompatible with an accepted categorization system [expressed in Murphy (1997) as a semantic field structure] either will not gain acceptance (and thus become obsolete) or will undergo semantic change in order to become compatible with the existing system.

Hypothesis 2 assumes that the verbal is secondary to the social and cognitive (belief structure) forces, that the mere introduction of new terminology cannot effect true social change. Hence, the limited success of *monosexual*, the semantic shifts in reclaimed *queer* (from an inclusive—homosexual, bisexual, transsexual, non-conformist heterosexual—to 'gay' with a racier name) and in *African American* (intended as an ethnic category on par with *Irish American*, but used as a synonym for *black* and contrast term for *white*).

3. Category names will be most closely associated with members of the category that are least like members of a relevant contrast category.

This is to say that social categories, like most categories, have a prototype structure. Because prototypes are formed with reference to contextually-relevant contrast categories, the meanings of socionyms are dynamic and subjective. For instance, the centre of a prototype-based category for *Catholic* will be construed differently if the relevant contrast category is *Protestant* or *Muslim* (Hogg/Terry/White 1995).

4. The more positively-valued the group, the more exclusive the senses and extensions of its name.

Tajfel and Forgas (1981: 121) frame this issue in terms of errors of over-inclusion and under-inclusion in assignment of others to a group, but the point holds generally for the categorization process, not just for errors. Thus, for instance, people with mixed racial heritage tend to be assigned to the more negatively-valued group (and hence the U.S. has its first black president, rather than its 44th white one).

5. People who identify within a category (and thus are likely to value that category positively in comparison to others) will use narrower senses of that category's name than will outgroup members.

This hypothesis is based on the observation that people have a general stereotype of their own group as good and deserving and of outgroups as bad and undeserving (Brewer 1979). If people are not free to re-categorize themselves because the category is ascribed to them on some "objective" grounds, or if the social and personal cost of re-categorizing is too great, they will find virtue in belonging to the category, regardless of its social standing. Thus, the hypothesis predicts that the association of positive value with exclusivity (as in hypothesis 4) is as valid for the "subjective" positive self-evaluation associated with group membership as it is for the "objective" positive evaluation determined by a high place on the social hierarchy. While there is some evidence that this happens, there are contrary forces, particularly the need for safety in numbers and the desire to promote the group to a 'normal' rather than 'negative' status, which work against this hypothesis. (For instance, lists of "[category members] in History" tend to err wildly on the side of over-inclusion, in an attempt to ameliorate perceptions of the category by associating it with many and high-status people. Thus Virginia Woolf, for example, ends up on lists of both lesbians and bisexuals in history.) Thus, we have two conflicting equations, MORE=BETTER and VALUABLE=RARE, whose aptness depends on the context and the category type—e.g. ascribed (by others) or affiliative. The conflict between these modes of evaluation, again, contributes to the fuzziness of socionym meaning.

Finally, it has also been observed that identification with negatively-valued (i.e. socially "marked") groups tends to be stronger than identification with positively-valued ("unmarked") groups and helps define the in-group in more socially secure terms (e.g. Meinhof 2004, in relation to East/West German identities and the negative identification of the ex-GDR people on the part of west Germans). Notice here, though, that this requires a positive formulation of the identity—a contrary rather than a contradiction, in Greimas's terms. While identities arise from the basic 'us/them', no one takes the identity 'them' or 'not us' for themselves. So, while *black* is only socially meaningful when it is in contrast to another category, usually *white*, it is only on rare occasions in the context of self-categorization that it is construed as having the privative 'not-white' sense. More typically *black* is associated with the having of qualities (a history, a phenotype, a set of cultural experiences) rather than the lack of whiteness.

In this section, we have discussed some ways in which a cognitive, social-psychological approach to identity categories can inform a discussion of the meanings of the terms associated with those categories. While some contextual inputs to asymmetry are socio-economic/historical (who has power, who has wealth), our deictic relations to social identity categories, that is, the shifting subjectivity of our perceptions of these categories, enforces asymmetries in terms of socionyms' polysemy, inclusiveness or exclusiveness, and positive or negative evaluation. These, in turn, can result in linguistic markedness asymmetries. The Social Identity Approach to social categorization gives us a means to form hypotheses about potential constraints on socionym meaning, which both acknowledges the general properties of categorization behaviour and posits additional special properties of social categorization.

CONCLUSION

Rather than providing answers, this chapter has pointed out ways in which modern linguistic research engages with the fact of asymmetrical semantic representation. While we question the definition of CONCEPT in the Koselleckian sense, we also see the value of treating some socio-political concepts (in the psychological sense) as special. In particular, the deictic nature of identity-based concepts gives rise to particular socio-cognitive constraints on the contextual construal of and semantic change in socionyms.

The concerns that we have expressed about the Koselleckian concept may not be new—several authors have noted that "Koselleck's metatheoretical statements [...] draw the fire of linguists" (Richter 1995: 50; also Rayner 1988). From our (admittedly anglocentric) linguistic perspective, we have not seen much of this fire—Koselleck's work has not been widely debated in our field. However it seems to us that Koselleck's historically determined socio-concepts are to an extent compatible with and can fruitfully dialogue with some recent developments in semantic and discourse studies. The connection between the discourse-analytic subdiscipline of linguistics and *Begriffsge-*

schichte are not hard to see (of course, discourse theories typically transcend disciplines). But we hope that we have also shown the worth of exploring the interface between the cognitive and the historical notions of CONCEPT and the potential for mutually interesting explorations, in which the practice of *Begriffsgeschichte* allows for testing of socio-cognitive-linguistic theories with reference to different historical periods.

REFERENCES

Althusser, L. 1971. *'Lenin and Philosophy' and Other Essays*. London: New Left Books.

Bally, C. 1940. L'Arbitraire du Signe. *Le Français Moderne* 8: 193-206.

Barsalou, L. W. 1983. *Ad* hoc Categories. *Memory and Cognition* 11: 211-227.

Bernstein, B. 1971. *Class, Codes and Control. Theoretical Studies Towards a Sociology of Language*. London: Routledge & Kegan Paul.

Benwell, B./Stokoe, E. 2006. *Discourse and Identity*. Edinburgh: Edinburgh University Press.

Brewer, M. B. 1979. In-group Bias in the Minimal Intergroup Situation. *Psychological Bulletin* 86: 307-323.

Bourdieu, P. 1992. *Language and Symbolic Power*. Cambridge: Polity Press.

Brown, P./Levinson, S. 1987. *Politeness: Some Universals in Language Use*. Cambridge: Cambridge University Press.

Coseriu, E./Geckeler, H. 1981. *Trends in Structural Semantics*. Tübingen: Narr.

Croft, W./Cruse. D. A. 2004. *Cognitive Linguistics*. Cambridge: Cambridge University Press.

Culpeper, J. 1996. Towards an Anatomy of Impoliteness. *Journal of Pragmatics* 25: 349-385.

Davies, M. 2007. The Attraction of Opposites. In *Stylistics and Social Cognition*, edited by L. Jeffries/D. McIntyre/D. Bousfield. Amsterdam: Rodopi, 79-100.

Dijk, T. van. 2008. *Discourse and Power*. Houndsmills, UK: Palgrave.

Dynes, W. 1985. *Homolexis. A Historical and Cultural Lexicon of Homosexuality*. New York: Scholarship Committee of the Gay Academic Union.

Fairclough, N. 1989. *Language and Power*. London: Longman.

Fairclough, N. 1992. *Discourse* and *Social Change*. Cambridge: Polity Press

Fellbaum, C. (editor). 1998. *WordNet. An Electronic Lexical Database*. Cambridge, MA: MIT Press.

Flexner, S. B. 1976. *I Hear America Talking*. New York: Van Nostrand Reinhold.

Foucault, M. 1972. *The Archaeology of Knowledge*. London: Tavistock.

Foucault, M. 1977. *Discipline and Punish*. New York: Pantheon.

Foucault, M. 1980. *Power/Knowledge*. New York: Pantheon

Gal, S./Irvine, J. T. 1995. The Boundaries of Languages and Disciplines. How Ideologies Construct Difference. *Social Research* 62: 967-1001.

Givón, T. 1970. Notes on the Semantic Structure of English Adjectives. *Language* 46: 816-837.

Givón, T. 1991. Markedness in Grammar. *Studies in Language* 15: 335-370.

Greimas, A.-J. 1983. *Structural Semantics. An Attempt at a Method* (translated by D. McDowell, R. Schleifer and A. Velie). Lincoln: University of Nebraska Press.

Hall, S. 1997. *Representation. Cultural Representations and Signifying Practices*. London: Sage.

Habermas, J. 1984. *Theory of Communicative Action. Volume 1*. London: Heinemann.

Handke, J. 1995. *The Structure of the Lexicon. Human versus Machine*. New York: Mouton de Gruyter.

Haspelmath, M. 2006. Against Markedness (and what to replace it with). *Journal of Linguistics* 42: 25-70.

Heins, V. 2004. Civil Society's Barbarisms. *European Journal of Social Theory* 7 (4): 499-517.

Hogg, M. A./Reid, S. A. 2006. Social Identity, Self-categorization, and the Communication of Group Norms. *Communication Theory* 16: 7-30.

Hogg, M. A./Terry, D. J./White, K. M. 1995. A Tale of Two Theories. A Critical Comparison of Identity Theory with Social Identity Theory. *Social Psychology Quarterly* 58 (4): 255-269.

Horn, L. R. 1984. A New Taxonomy for Pragmatic Inference. Q-based and R-based Implicature. In *Meaning, Form and Use in Context,* edited by D. Schiffrin. Washington, DC: Georgetown University Press, 11-42.

Horn, L. R. 2001. *A Natural History of Negation. Revised Edition.* Stanford, CA: CSLI.

Høstaker, R. 2005. Latour. Semiotics and Science Studies. *Science Studies* 18 (2): 5-25.

Israel, M. 2004. The Pragmatics of Polarity. In *The Handbook of* Pragmatics, edited by L. R. Horn/G. L. Ward. Oxford: Blackwell, 701-723.

Jackendoff, R. 2006. On Conceptual Semantics. *Intercultural Pragmatics* 3: 353-358.

Jeffries, L. 2010. *Opposition in Discourse.* London: Continuum.

Jones, S. 2002. *Antonymy*. London: Routledge.

Jones, S. 2007. 'Opposites' in Discourse. *Journal of Pragmatics* 39: 1105-1119.

Jones, S./Murphy, M. L./Paradis, C./Willners, C. forthcoming. *Antonymy in English.* Cambridge: Cambridge University Press.

Kaufer, D. 1981. Understanding Ironic Communication. *Journal of Pragmatics* 5: 495-510.

KhosraviNik, M. 2010. The Representation of Refugees, Asylum Seekers and Immigrants in British Newspapers. *Journal of Language and Politics* 9: 1-28.

Koselleck, R. 2004. *Futures Past. On the Semantics of Historical Time* (translated by Keith Tribe). New York: Columbia University Press.

Kripke, S. 1980. *Naming and Necessity*. Cambridge, MA: Harvard University Press.

Krzyzanowski, M./Wodak, R. 2008. *The Politics of Exclusion. Debating Migration in Austria.* New Brunswick: Transaction Press.

Lakoff, G./Johnson, M. 1980. *Metaphors We Live By*. Chicago: University of Chicago Press.

Landaur, T. K./Dumais, S. T. 1997. A Solution to Plato's problem: The Latent Semantic Analysis Theory of Acquisition, Induction, and Representation of Knowledge. *Psychological Review* 104: 211-240.

Leeuwen, T. van. 2008. *Discourse and Practice*. Oxford: Oxford University Press.

Levinson, S. C. 2000. *Presumptive Meanings. The Theory of Generalized Conversational Implicature*. Cambridge, MA: MIT.

Lipson, M. 2009. 'If it wasn't rolling, it never happened': The Role of Visual Elements in Television News. In *Evaluation and Stance in War News*, edited by L. Haarman/L. Lombardo. London: Continuum, 140-169.

Ljung, M. 1974. Some Remarks on Antonymy. *Language* 50: 74-88.

Malkiel, Y. 1959. Studies in Irreversible Binomials. *Lingua* 8: 113-160.

Marchi, A./Taylor, C. 2009. Who was Fighting and who/what was being fought? The Construction of Participants' Identities in the UK and the US Reporting of the Iraq War. In *Identities across Media and Modes. Discursive Perspectives*, edited by G. Garzone/P. Catenaccio. Bern: Peter Lang, 259-287.

Martin, J. 1999. Grace. The Logogenesis of Freedom. *Discourse Studies* 1: 29-56.

McDonald, J. L./Bock, K./Kelly, M. H. 1993. Word and World Order. Semantic, Phonological and Metrical Determinants of Serial Position. *Cognitive Psychology* 25: 188-230.

Meinhof, U. H. 2004. Europe Viewed from Below. Agents, Victims, and the Threat of the Other. In *Transnational Identities: Becoming European in the EU*, edited by R. Herrmann/T. Risse/M. Brewer. Lanham, MD: Rowman & Littlefield Publishers, 214-244.

Miller, G. A. (editor). 1990. *WordNet. An on-line lexical database. Special issue of International Journal of Lexicography* 3 (4).

Mills, S. 2004. *Discourse*. London: Routledge.

Muehleisen, V./Isono, M. 2009. Antonymous Adjectives in Japanese Discourse. *Journal of Pragmatics* 41: 2185-2203.

Murphy, G. L. 2002. *The Big Book of Concepts*. Cambridge, MA: MIT Press.

Murphy, M. L. 1991. Defining Racial Labels. Problems and Promise in American Dictionaries. *Dictionaries* 13: 43-64.

Murphy, M. L. 1997. The Elusive Bisexual. Social Categorization and Lexico-semantic Change. In *Queerly Phrased: Language, Gender, and Sexuality*, edited by K. Hall/A. Livia. Oxford: Oxford University Press, 35-57.

Murphy, M. L. 1998. Defining People. Race and Ethnicity in South African English Dictionaries. *International Journal of Lexicography* 11: 1-33.

Murphy, M. L. 2003. *Semantic Relations and the Lexicon.* Cambridge: Cambridge University Press.

Murphy, M. L. 2006. Antonyms as Lexical Constructions. Or, Why *Paradigmatic Construction is Not an Oxymoron*. In *Constructions*, SV1–8/2006. http://www.constructions-online.de/articles/specvol1.

Murphy, M. L./Jones, S. 2008. Antonymy in Children's and Child-directed Speech. *First Language* 28: 403-430.

Murphy, M. L./Paradis, C./Willners, C./Jones, S. 2009. Discourse Functions of Antonymy. A Cross-linguistic Investigation of Swedish and English. *Journal of Pragmatics* 41: 2159-2184.

Outlaw, L. 1990. Toward a Critical Theory of 'Race'. In *Anatomy of Racism*, edited by D. T. Goldberg. Minneapolis: University of Minneapolis Press, 58-82.

Paradis, C./Willners, C. 2006. Antonyms and Negation. The Boundedness Hypothesis. *Journal of Pragmatics* 38: 1051-1080.

Piazza, R. 2009. News is Reporting What was Said. Techniques and Patterns of Attribution. In *Evaluation and Stance in War News*, edited by L. Haarman/L. Lombardo. London: Continuum, 170-194.

Piazza, R. 2010. *Let Cinema Speak. The Discourse of Italian Cinema and Beyond*. London: Continuum.

Poole, E. 2002. *Reporting Islam. Media Representations of British Muslims*. London: Tauris.

Rayner, J. 1988. On *Begriffgeschichte*. *Political Theory* 16: 496-501.

Richter, M. 1995. *The History of Political and Social Concepts*. Oxford: Oxford University Press.

Said, E. [1978] 2003. *Orientalism*. London: Penguin.

Sampson, G. 2000. Review of WordNet. An Electronic Lexical Database. *International Journal of Lexicography* 13: 54-59.

Saussure, F. de. [1915] 1959. *Course in General Linguistics*. London: Peter Owen.

Searle, J. [1979] 1991. Metaphor. In *Pragmatics. A Reader*, edited by S. Davis. Oxford: Oxford University Press, 519-539.

Sperber, D. 1984. Verbal Irony. Pretense or Echoic Mention? *Journal of Experimental Psychology: General.* 113: 130-136.

Sperber, D./Wilson, D. 1981. Irony and the Use-Mention Distinction. In *Pragmatics. A Reader*, edited by S. Davis. Oxford: Oxford University Press, 550-564.

Sweetser, E. 1990. *From Etymology to Pragmatics*. Cambridge: Cambridge University Press.

Tajfel, H. 1981. Social Stereotypes and Social Groups. In *Intergroup Behaviour*, edited by J. C. Turner/H. Giles. Oxford: Blackwell, 144-167.

Tajfel, H./Forgas, J. P. 1981. Social Categorization. Cognitions, Values, and Groups. In *Social Cognition*, edited by J. P. Forgas. New York: Academic, 113-140.

Tajfel, H./Turner, J. C. 1979. An Integrative Theory of Intergroup Conflict. In The *Social Psychology of Intergroup Relations*, edited by W. G. Austin/S. Worchel. Monterey, CA: BrooksCole, 33-47.

Thornborrow, J. 2002. *Power Talk. Language and Interaction in Institutional Discourse*. London: Longman

Trubetzkoy, N. 1939. *Grundzüge der Phonologie*. Prague: Cercle Linguistique de Prague.

Turner, J. C. 1984. Social Categorization and the Self-concept. A Social Cognitive Theory of Group Behavior. *Advances in Group Processes. Theory and Research* 2: 77-122.

Vendler, Z. 1963. *The Transformational Grammar of English Adjectives*. Philadelphia: Department of Linguistics, University of Pennsylvania.

Wierzbicka, A. 1997. *Understanding Cultures through their Key Words*. Oxford: Oxford University Press.

Wierzbicka, A. 2006. *English. Meaning and Culture*. Oxford: Oxford University Press.

Williams, G. 2007. Europe in *Le Monde* 2004. *Mediazioni. Online Journal of Interdisciplinary Studies on Languages and Cultures*, University of Bologna. http://www.mediazionionline.it/.

Williams, G./Piazza, R./Giuliani, D. In press. Nation and Supernation: A Tale of Three Europes. In *European Identity. What the Media Say*, edited by G. Williams/P. Bayley. Oxford: Oxford University Press.

Wittgenstein, L. 1958. *Philosophical Investigations. 3rd Edition* (translated by G. E. M. Anscombe). New York: Macmillan.

Wodak, R. 1989. *Language, Power and Ideology*. Amsterdam: Benjamins.

Wodak, R. 2004. The Discourse-historical Approach. In *Methods of Critical Discourse Analysis*, edited by R. Wodak/M. Meyer, London: Sage, 63-94.

Wodak, R. 2009. *The Discourse of Politics in Action. Politics as Usual*. Basingstoke, UK: Palgrave.

Wodak, R./Reisigl, M. 2003. Discourse and Racism. In *The Handbook of Discourse Analysis*, edited by D. Schiffrin/D. Tannen/H. E. Hamilton. Malden, MA: Blackwell, 372-397.

Yariv-Laor, L./Sovran, T. 1998. The Structure of Linguistic Asymmetry: Evidence from Hebrew and Chinese. *Papers and Studies in Contrastive Linguistics* 34: 199-213.

Asymmetrical Concepts and Political Asymmetries

A Comparative Glance at 20th Century Democracies and Totalitarianisms from a Discursive Standpoint*

KIRILL POSTOUTENKO

Strange as it may seem, the so-called asymmetrical concepts, discovered some 35 years ago by Reinhard Koselleck (Koselleck 1975) and promoted since then to the core of "conceptual history", have yet to see the proof of their importance in small-and middle-scale empirical studies. However, this vacuum appears less surprising given the fact that even the founding fathers of *Begriffsgeschichte* were perplexed when it came to anchoring asymmetrical concepts in social reality (Berding 1976: 110). In fact, lamenting the restlessness of conceptual history within modern social sciences and humanities became lately a sort of a commonplace (see, for example, Gumbrecht 2006: 28). Together with the other studies collected in this volume, the present paper strives to offer an alternative to this somewhat fruitless pessimism. Firstly, the original conceptual scheme of Reinhart Koselleck (I) is summed up, checked for inconsistencies and omissions and updated accordingly. Secondly, asymmetrical concepts are shown to be parts of the bigger asymmetries governing the relations between dissipative

* The work on this article was supported by the Institute for Advanced Studies (Paris, France).

systems in nature and society (II).[1] Finally, the empirical study of conceptual asymmetries in the political discourse of the "democratic" (Franklin D. Roosevelt) and "totalitarian" (Adolf Hitler and Joseph Stalin) leaders, establishes correlation between large- and small-scale asymmetries (III): the original, rather crude hypothesis linking societies with greater communicative asymmetries (Soviet Union and Nazi Deutschland) to the texts with the stronger presence of asymmetrical concepts, is subsequently refined with the help of the data obtained in the course of statistical examination.

I

Seeking to explain the emergence and role of "conceptual oppositions" (*Gegenbegriffe*) in European culture, Reinhard Koselleck formulated (on various occasions) several heterogeneous hypotheses, which can be reduced to the following three lemmas:

1. There is a number of psychological, economical, religious, social, geographical and political circumstances capable of generating interpersonal animosities. Each social actor able and willing to turn this unfocused sensation of hostility into a full-blown distinction between a *Friend* and an *Enemy* may draw upon preexisting differences between dialects and jargons (Koselleck 1994: 275, 280-81).

2. The opposition between "*Hellenes*" and "*Barbarians*" is rooted first and foremost in territorial difference between *inner* and *outer* positions. The inner position, immediately attached to the *Self*, is invariable: "Hellenes" always remain Hellenes. Conversely, the outer, variable position speculatively projected onto the distant *Other*, may very well vary: the role of "Barbarians" has been successfully played at times by Persians, Scythes and Galls (Koselleck 1994: 276; Koselleck 2003: 104).[2]

---

1 The full version of this part is also presented below as a separate article. Although every effort has been made to avoid redundancy, some overlapping seemed indispensable for making texts readable independently from each other.

2 In the early, more radical version, *both* self- and other-terms are paradigms of rotating signs which arbitrariness is only constrained by the necessity of

3. "Conceptual oppositions" are indispensable for spelling out identities as differences ('*Us*' vs. '*Them*') (Koselleck 1994: 276).

For the most part, these postulates leave little room for disagreement, but the overall impression they make is that of a jigsaw puzzle with a lot of pieces missing. Worse, the pieces that are actually on hand point at different, sometimes incompatible research traditions housed in various disciplines: small wonder that integrating conceptual history into general social theory has proved to be such a difficult task. Any serious attempt at such integration would invariably consist not only of bridging gaps between scattered theses, but also of streamlining equivocal formulations and even getting rid of some "foreign bodies" incompatible with the rest of the argument.

The first Koselleck's thesis paints the gloomy picture of *bellum omnium contra omnes* (Hobbes 1647: 148), which is expected to be a fertile ground for all sorts of interpersonal and intergroup differentiations. According to this view, human universe starts playing by the rules of structuralist utopia and political metaphysics as soon as some politically active individuals begin transforming potential differentiability of human race into social reality: little by little, social life gets organized into binary oppositions in the flesh—the pairs *friend / enemy* and their derivatives (Ivanov 1974; Ivanov 1983: 37; Schmitt 1932: 35-36, 105).[3] Many aspects of this reconstruction look probable, but neither the roots of the lamentable *status quo* nor the causal link between initial chaos and resulting order are spelled out clearly enough.

To remove the first uncertainty, it may suffice to recast it in terms of modern science. Being an open dissipative system, the world invariably consists of objects with different energetic (and informational) value (Horwitch 1987: 72). Under lasting equilibrium conditions, social communication would come to a full stop or never begin in the first place, since there would be nothing to communicate and, consequently, little reason to engage in interaction (Shannon/Weaver 1949:

discernible semantic difference within a pair: naturally, "Hellenes" cannot be opposed to "Hellenes" (Koselleck 1975: 216; see also Günther 1979: 103).

3 Schmitt's influence on Koseleck has recently become an object of detailed research (see Olsen 2004; Missfelder 2006; Villas Boas Castello Branco 2006; Pankakoski 2010).

13). This, by the way, explains the general toleration of informational asymmetries in all sorts of communicative settings from private conversations to marketplaces: the elimination of informational disparities between individuals and small groups, except for being exceedingly costly, could hardly have had but a short-term effect (Lehmann 1999: 10).

The second ambiguity, on the contrary, cannot be removed as easily, because there is no immediate explanation of how various individuals with different private knowledge would all of a sudden divide themselves into pairs and form perfect privative oppositions. Apparently, the only thinkable reason for intentional pairing is communication (Luhmann 1987: 199): the person calling his or her conversation partner '*Enemy*' expects, regardless of rage and indignation, to be heard and somehow understood. This, in turn, presupposes some degree of symmetrical closeness between interaction partners in interpreting the code in question: no matter how big the difference of opinions and temperaments is, there must be some agreement between conversation partners on what the sequence of graphemes *e*, *n*, *e*, *m* and *y* might mean, because otherwise no coded interaction is possible. In contrast, the more advanced symmetries of lexical meanings in contradictory or privative oppositions are much less regular than the breakdown of each and every speech act into a sender and a recipient, and this irregularity seems to disqualify them for the constitutive role in micro-level social differentiation envisioned by Koselleck. This is not to deny that for each "asymmetrical concept" (for example, "*ridiculous*", "*mad dogs*" or "*poking their noses into…*") there is a cluster of lexemes capable of serving as its opposites in a given context (for example, "*reasonable*"/"*serious*"/"*thought-out*", "*humans*"/"*sane living beings*" or "*minding one's own business*"/"*leaving others alone*"). But the vague semantic asymmetry between "Hellenes" and "Barbarians" plays quite a limited role in the meaning of the latter term compared to the strong perceptual asymmetry at both ends of interaction: indeed, defining '*Barbarians*' solely as anti-'*Hellenes*' (and vice versa) would be a big stretch on all counts; on the other hand, nobody has ever seriously expected from "Barbarians" to apply this derisive and absurdly uninformed name to themselves (Luhmann 1995: 140).[4] Apparently,

4 This is not to deny that the identity terms reserved for signifying Otherness having been used reflexively on various occasions. Already Michel de

an asymmetrical relation does not unite—or disunite concepts like '*Friends*' and '*Enemies*': rather, it sets off the *sender* of the latter message against some of its *recipients*.[5] This fact allows for reassessing one of the axioms of *Begriffsgeschichte*: far from being conceptual pairs observed from a single perspective, asymmetrical concepts turn out to be single utterances looked at from two different angles.[6]

Montaigne took an issue with the rebirth of Hellenic egocentricity in emerging Western Enlightenment: "Nous les pouvons donc bien appeler barbares, eu esgard aux regles de la raison, mais non pa eu esgard à nous, qui les surpassons en toute sorte de barbarie" (Montaigne 1595: 242). But this trend has nothing to do with restoring conceptual symmetry because the very semantics of such and similar attributive inversions thrives on their marginality in relation to the original opposition. An honest attempt at such restoration has been made by Herodotus in the opening lines of his *Histories* (450-420 BC) which very syntax, as symmetrical as it could be, hints at the (positive) equality of both parties: those masterpieces which sliding into oblivion the founder of historical science tries to arrest, are said to be brought into being by Hellenes and Barbarians alike (τά μέν λλησι τά δέ βαρβάροισι ἀποδεχθέντα (Herod. *Historiae*, A 1-5)). Evidently, Herodotus' impartiality has found throughout the centuries little resonance.

5 The word "some" is used to prevent simplifications: a recipient may turn out to be another "Hellene" who feels about the "Barbarians" the same way as the speaker does. Later it will be shown that even in a totalitarian world where "Others" are extinct or pretends to be "Selves", there is still room for inalienable asymmetries in each and every interpersonal relation imaginable.

6 To be sure, this reassessment would be unthinkable without pragmatic and socio-interactionist revision of conceptual history which is taking place on its margins for quite a while (see Schultz 1979; Busse 1987: 304; Hass 1991: 331, 335; Knobloch 1992: 11; Gloy 1996: 360; Guilhamou 2000; Palonen 2002: 229; Carlson 2003; di Sciullo 2003: 1; Lübbe 2003). Already the classical definition of "essentially contested concepts" ('*democracy*', '*justice*' etc.) contains the program of reorganizing *Begriffsgeschichte* on inter-subjective grounds: "To use an essentially contested concept means to use it against other uses and recognize that one's own use of it has to be maintained against these other uses" (Gallie 1956: 172).

The relegation of asymmetry from lexical semantics to conversational pragmatics results in a smoother transition from Hobbesian warfare to the socially cohesive universe: it seems reasonable that cognitive inequalities and evaluative disparities *between* individuals *X* and *Y* cause interpretative asymmetry *between* the statement *n* uttered by *X* and the same statement *n* perceived by *Y* (or vice versa). What still remains to be seen are the origins of asymmetric advantage possessed, *in potentia*, by each and every speaker. Koselleck's favorite example seems to hint at the differences between regional and social discourses as *the* breeding ground for conceptual asymmetries: by derisively referring to the speechless mutterers (τουσ βαρβαρουσ), the prototypical Hellenic speaker masterfully sells his own linguistic deficiency as his linguistic advantage over his counterpart. But it stands to reason that in actual discursive behavior very few conceptual asymmetries boast such a fanciful design, no matter how popular it is in post-structuralist philosophy and post-imperialist cultural criticism (see Lyotard 1983: 9; Said 1979: 166-198). Indeed, making a virtue out of ignorance is rarely possible without silencing (or at least ignoring) dissenting voices, so if the word Barbarians is to make sense, there must be ways to deny them common ground in any real or imaginary interaction—either by disqualifying them from communication altogether ("*those speechless barbarians cannot but mumble*") or by treating their communicative qualifications as irrelevant under conditions dictated by the speaker ("*even if barbarians can speak, their speech has no meaning since we cannot understand it*"). Evidently, none of these denials makes linguistic sense and could be accepted by an unprejudiced observer armed with the average knowledge of Aristotelian logic: rather, each of the two sentences looks like a guise for a high-order asymmetry, expressed—accidentally—through discourse but in reality independent from it.

II

Koselleck's second and third lemmas shed some light on this overarching asymmetry, which most consistent explanation—again—could be given in terms of systems theory. As the preference given to self-organization—as opposed to integration with the environment—is indispensable for the intactness of every living system, it seems natural for human life to hinge on the two intertwined forms of self-identification: defining the set of systemic parts (as opposed to alien objects) goes hand in hand with maintaining totality of their relations essential for the system's survival (as opposed to optional relations, including relations with alien objects) (see Zeleny 1980: 5). The first identifying distinction (*Self* vs. *Other*) is often associated with the activity of the immune system (Jacob 1982: 54; Tauber 1994), but already at the most basic level of life one could with similar ease detect the primal asymmetry of parts (chiral molecules) and their relations (non-racemic mixtures of "left" and "right" enantiomers) (Neville 1976: 3). The second distinction (*Inside* vs. *Outside*), in its turn, inscribes this asymmetry into the centralized architecture of systemic self-control: the very architecture of nervous system, broken down into a heavily protected center and a spreading periphery extending to the points of contact with the environment (Churchland 2003), defines the concentric shape of human identity in social life, confirmed by the very notion of "personal space" (Sommer 1969): the *Self* ('*I*'), immediately felt and invariably important, is located *Inside* ('*Here*'), whereas the *Other* ('*You*'), seen at a distance and potentially harmful, is placed *Outside* ('*There*') (Butterworth 1995). At this point it is abundantly clear that Koselleck's second and third theses override the first one: alone the stark contrast between the tightly knit pair of "positive" self-referential terms '*I*'/'*Here*' and its "negative" reflection '*You*'/'*There*' which can barely hold together, suffices to reconfirm that "Barbarians" could be anything but Hellenes—but not vice versa.[7] But it still remains to be specified how this immediate feeling of "being inside" and invariable importance of "self-control" are being converted into conceptual asymmetries that place the denotate of '*I*' and '*Here*' head and shoulders above the signified of '*You*' and '*There*'.

The answer to this question might perhaps be simpler than one thinks. The fact that the self-referential (pre-social) part of human identity is a heavily centralized system which subsystems manage qualitatively homogeneous matter (*Self*) in a nearly exclusive domain (*Inside*), results in a number of reflexive presuppositions pertaining to

---

7 The empirical evidence supporting this supposition could be found in the language of Soviet propaganda where the "good boys" are associated with *totality* and *continuity* and the "bad boys" are linked to *partiality* and *transience* (see Weiss 2000: 240).

the individual's spatial centrality (‘*I am the starting point for the surrounding environment*’), privileged access to own states (‘*I know my feelings better than others*’), superiority of beliefs (‘*I have better reasons to believe in my truth than others have*’) and, last but not least, the higher degree of self-organization compared to the surrounding environment (‘*I am more stable than other systems*’) (see details in Davidson 2001: 38; Wittgenstein 1951: 24; Tugendhat 1979: 33; Chilton 2004: 59). It seems quite natural that verbal communication provides abundant means for accommodating such interpersonal asymmetries, which usually entails their minimization: the success of the speech act between *X* (sending a message ‘*I feel pleasure*’ to *Y)* and *Y* (receiving it) requires that both interlocutors should have at least some shared understanding of ‘pleasure’, otherwise their asymmetry would outgrow communicative realm and simultaneously lose connection with the perceptual content of *X*’s message—its privileged access to its own states. However, the very existence of conceptual asymmetries shows that shared understanding in communication has its limits. In the given example, for instance, *X* and *Y* cannot interpret the word ‘*pleasure*’ in the same way, because, among other things, there is no rationally imaginable *sender/recipient* pair which observational standpoints in regard to the sentence in question could be identical: even if X and Y both “feel pleasure” (say, riding a merry-go-round), the feeling *X* communicates to *Y* in a sentence ‘*I feel pleasure*’ is *not* the pleasure *Y* feels at that moment, or any moment before or after. Even a handful of asymmetrical concepts discussed so far reveals remarkable variability of orientational and observational stances embedded in their meanings: being, in all probability, neither accidental nor purely situational, such variations seem to follow semiotic (and other communicative) hierarchies within human language.

Tying the classification of conceptual asymmetries to these hierarchical structures appears to be practical for two reasons. Firstly, as far as general classification of asymmetrical concepts is concerned, semiotics appears to be a reliable quantitative indicator, capable of locating asymmetries in question on a sliding scale between the raw Hobbesian “action” as a whole and its subtler subsystems specialized on long-term interpersonal coordination (such as communication). Thus, it seems clear that asymmetry is far stronger in a message with a high *indexical* content than in a message conveyed entirely in *symbols*: one may compare the divisiveness of the statement ‘*You are invader*’ which last word unequivocally points at the limits of the speaker's personal or group space, with the much less contentious sentence '*You are racing*' containing no such reference. Furthermore, in the first statement the asymmetrical content of the predicate ('*an unwelcome outsider*') is tightly knit with the sender's implicit reflexive reference to its own living being (['*I am the insider* (*who decides who is welcome and who is not*')]) and cannot therefore—in its present form—lead to shared understanding between the *current* speaker and the referent. But in the second case such a "symmetrical" consensus appears to be more likely to achieve, since the asymmetrical meaning of the term in question ('*racing*' = '*moving faster than I consider appropriate*') may be fairly easily decoupled from speaker's position and orientation and "universalized" through a mutual reference to a social norm shared by communication partners ('*racing*' = '*going faster than 60 mph*').

Secondly, this small-scale hierarchy of semiotic differentiation within asymmetrical concepts has direct large-scale analogies on the societal level: for instance, the differences between action and utterance (which constitute *signs* in the first place) and between message and information (which produces *symbols* such as words), being common for advanced communicative systems (Luhmann 1987: 193-195), may be partially suspended in authoritarian societies, which interactional practices on many occasions imply the ritualistic unity of body, its communicative role and its political power (Leach 1976: 37; Rappaport 1999: 58; Postoutenko 2010a: 27-28). Discouraging exchange of political and communicative roles, such a suspension inevitably leads to a vast social asymmetry (Habermas 1974: 264): the discourse in which various references to "*the speaker comrade Stalin*" nearly outnumber the references to all other *speakers*, *human beings* and *state officials* combined—the case of the newspaper *Pravda* during the 8th Extraordinary Congress of Soviets in November 1936—inevitably suggests that there is only one living being in the world capable of giving speeches and performing executive functions. As the inquiry into functioning of asymmetrical concepts in public communication is based in this article on comparing and contrasting conceptual asymmetries in "totalitarian" and "democratic" systems, this analogy between micro- and macro-semiotics may be instrumental in making this multilevel comparison of communicative asymmetries consistent and free from ideological labels of the past.

III

In this vein, it seems pertinent to suggest a unified classification of conceptual asymmetries in terms suitable for empirical analysis. Departing—as has been suggested before—from Peirce's semiotic triad, one may first of all draw the line between asymmetrical concepts which semiotic differentiation is still at its formative stages (*index* pointing at the sender and *symbol* referring to the potentially sharable information form a semiotic unity) and the ones where *symbolic* component is more independent and could be to some extent perceived independently in the appropriate semantic and interactional context. The first category, rooted in the special *position* of human system in relation to environment (and best illustrated by the basic deictic triangle '*I*'-'*here*'-'*now*'), could be called *primary (orientational) asymmetries*, whereas the second, predominantly based on preferential *access* of human system to its own states and beliefs (and resembling the vaguer triad (['*I*'] (hiding behind the curtain)-'*pleasant*'-'*right*') could be called *secondary (evaluative) asymmetries*. It has been already suggested that primary asymmetries are more robust than the secondary ones, because their speaker-centered egocentricity could be attenuated a bit but never eliminated in its entirety: the speaker may not be very prominent in '*we*', but there is no way he or she could be absent in '*here*' or non-existent in '*now*'. Nevertheless, the actual impact of this robustness on the large-scale social asymmetries is dampened by two parallel subsystems within verbal language responsible for equilibration of social semantics. On the one hand, primary asymmetries can be arranged into nearly perfect symmetrical structures by redirecting the reflexivity of their meanings from individual beings to the grammatical categories: indeed, if the *indexical* pointing at this particular sender in '*here*' or '*now*' is reduced to an empty sign (the reference to *some* utterer of the message), then the *symbolic* components of the respective meanings read like entries in an explanatory dictionary—the standard references to lexical norms. On the other hand, the orientational bias of primary asymmetrical concepts can usually be overridden by the *intrinsic* (as opposed to *deictic*) layer of language, largely based on symbolic codes with negligible social referentiality (such as logic or mathematics) and therefore free from indexical ballast in the first place: in this sense, '*here*' can be effortlessly translated into '*in*

Paris' or 'at 69° 40' N', and '*now*' would become '*in winter*' or '*at 5 p.m*' (Postoutenko 2007: 109-110).

Obviously, the same mechanisms of symmetrization are at work in the realm of secondary asymmetries, but their efficiency is significantly lower: unlike relatively simple and externally graspable *positional disparities* between *my* "here" and *your* "here", the *perceptual differences* between *my* "pain" and *your* "pain" may fill volumes. As a result, the systemic unity of lexical semantics fails to prevent or even offset the infiltration of interactional sphere with the feelings, beliefs and evaluations of a single communicator: in principle, each sender has at its disposal the means of reducing the impact of conceptual asymmetry upon the recipient—for example, forewarning ('*In my opinion*, this decision is wrong')—but the cases of such unpractical altruism are few and far between.[8] All things considered, the crucial role of evaluative asymmetries in upholding large-scale social asymmetries seems to be very plausible (although not ascertained at this point), and this plausibility dictates the specifics of data selection for this exploratory study: whereas the analysis of *primary asymmetries* is postponed until a later, more comprehensive study,[9] the *secondary asymmetries* are given a closer look and broken down into subcategories which relative strengths are derived from the level at which they affect communicative systems: the higher this level is, the stronger asymmetries are hypothesized to be. Besides expanding Koselleck's tripartite classification '*Hellenes*' vs. '*Barbarians*'/'*Christians*' vs. '*Pagans*'/'*Übermenschen*' vs. '*Untermenschen*' (see Koselleck 1975: 213) and providing stricter differentiation criteria, such a gradation seems to be convenient for seeking micro-macro correlations of social asymmetries outlined above. In conformity with this task, the classification below is built upon *communicative* rather than *linguistic* foundations. The sole *principium divisionis* between the categories is the opportunity of conveying information in a model interactional setting (*sender* → *recipi-*

8 In the empirical analysis of conceptual asymmetries (below), the utterances containing such explicit forewarnings were seen as non-asymmetrical regardless of their verbal content.

9 In fact, the scholars dealing with the semantics of personal pronouns have already come up with a number of insights concerning the social functions of primary conceptual asymmetries (see von Wiese 1965; Mühlhäusler/Harré/Freyne 1990; Postoutenko 2009).

ent), presented in the increasing order, whereas grammatical or semantic features of utterances belonging to the same category may be as heterogeneous as possible.

As far as secondary (evaluative) asymmetrical concepts go, the most obvious differences lie between *antagonistic asymmetries* (which semantic, syntactic or pragmatic properties block symmetrical reinterpretation from the start) and *agonistic asymmetries* (which meanings do leave room for such a reinterpretation even within the given interactional setting). Furthermore, as the width of the gap between the speaker and the recipient varies not only between two kinds of asymmetries but also within them, it seems practical to suggest a further breakdown of each category into *strong* or *weak* variations and then align four subcategories into a single descending scale of communicative complexity.

Strong antagonistic asymmetry presents the most radical challenge to shared understanding between interaction partners because it grants the sender the *monopoly on communication* and, consequently, denies a recipient the possession of general systemic properties (such as human identity) indispensable for symbolic interaction: following this logic, the ability of '*beasts*', '*capitalist sharks*' or other possessors of '*primitive instincts*' deserving to be '*slaughtered*', to *evaluate* surrounding environment cannot even be discussed seriously.[10] The positive variation of this asymmetry puts the same kind of a wall between the recipient and the world in question in a more paradoxical way: as the minimal level of communicative competence is raised above the

10 See examples: "Wer glaubt, im Interessenkreis dieser Freiheit und dieses Lebens nicht bestehen zu können, *hat keine Existenzberechtigung in unserer Gemeinschaft*" (Hitler 1935: 43); "For twelve years this Nation was *afflicted* with hear-nothing, see-nothing, do-nothing Government" (Roosevelt 1936m: 568); "Мой совет, совет кандидата в депутаты своим избирателям – помнить об этом праве избирателей – о праве досрочного отзыва депутатов, следить за своими депутатами, контролировать их, и, ежели они вздумают свернуть с правильной дороги, *смахнуть их с плеч*, потребовать назначения новых выборов" (Stalin 1937a: 240).

level of social normalcy, being "just a human" would not buy a living being admission to the negotiation table with '*Übermenschen*'.[11]

In contrast, the *weak antagonistic asymmetry* in principle allows a recipient to be a part of communication process, but reduces his or her role to that of a passive observer. The fusion of explicit egocentric orientation with the *monopoly on evaluation* in ethical ('*rightness*'/ '*unjust*'),[12] deontic ('*necessary*'/'*inacceptable*'),[13] epistemic ('*true*'~ '*reality*'~'*evidently*')/'*liars*'),[14] communicative ('*understandable* [sender's message]'/'[recipients] *unable to grasp*')[15] or unspecified *a pri-*

---

11 In the material analyzed below and presented in the *Tables 1-3*, the overwhelming majority of the *strong antagonistic asymmetries* were on the negative side. Giving exact breakdown would have made little sense as the figures were too small to justify any meaningful interpretation.

12 See examples: "Noch stets hat *die Richtigkeit* unserer Lehre und Grundsätze bei diesen größten Demonstrationen der Bewegung Ihre stärkste Bekräftigung erhalten durch den immer wieder sichtbaren Erfolg" (Hitler 1936: 30); "It still preaches the same *heresy*—class against class and religion against religion" (Roosevelt 1936x: 551); "Центральный комитет ударил по этим *ошибкам*" (Stalin 1937b: 196).

13 See examples: "Die deutsche Regierung wird daher die sonstigen, das Zusammenleben der Nationen betreffenden Artikel, einschließlich der territorialen Bestimmungen, unbedingt respektieren und die im Wandel der Zeiten *unvermeidlichen* Revisionen nur auf dem Wege einer friedlichen Verständigung durchführen" (Hitler 1935: 41); "For many years he went about the unspectacular but very *necessary* job of clearing the ground for the progress of a great labor movement" (Roosevelt 1936h: 547); "*Нельзя* на основании речей, деклараций, словесных заявлений делать вывод о природе, так сказать, данного работника" (Stalin 1937b:194).

14 See examples: "Es ist auch den jüdischen Berichterstattern nicht mehr möglich, diese *Wahrheit zu verdrehen* [1] und das, was Millionen mit eigenen Augen zu sehen und nachzuprüfen in der Lage waren und sind, *in das Gegenteil umzulügen* [2]!" (Hitler 1936: 35); "It's a wise and *truly* American policy" (Roosevelt 1936d: 538-539) "*Факты*, однако показали, что скептиками опять не повезло: пятилетний план был осуществлен в четыре года" (Stalin 1936: 27).

15 See examples: "Nicht wir sind für die Ursachen dieser Sorgen verantwortlich, sondern eine Schicksalhafte Vergangenheit sowie das leider oft so große *Unverständnis* [1] unseres eigenen Volkes und das *Nichtverstehen-*

ori (‘*ideal*’/‘*enemy*’)[16] terms makes the meanings of such utterances as ‘*immoral*’, ‘*unique*’ or ‘*best political system possible*’—for all their dependence on the recipient’s formal acceptance—*de facto* incontestable. (In contrast, the dogmatic assertion or negation of ‘beauty’ does not rule out deliberation, since the meanings of extremes in aesthetic evaluations resist being arranged into oppositions: “If the beautiful excludes the ugly, beauty is no measure of aesthetic merit; but if the beautiful may be ugly, then ‘beauty’ becomes only an alternative and misleading word for aesthetic merit” (Goodman 1968: 255)). Since the terms employed in weak antagonistic asymmetries are not gradable—normally, being everything else than ‘*right*’ is considered ‘*wrong*’—the common ground between two sides of asymmetry is limited to shared codes (verbal language) and practices (turn-taking) as such, and does not extend to the substantial correlation of communicators’ interpretative positions. Simply speaking, the only cooperative behavior at the recipient’s disposal is to agree, since the dual logic, inherent in the semantics of weak antagonistic asymmetries, would turn *any* disagreement into a deadlock (‘this statement *true*’→ ‘*no* [, this statement is *false*]’) unsolvable by communicative means.

---

können [2] und *-wollen* [3] anderer” (Hitler 1936: 36); “Our resplendent economic aristocracy does not want to return to that individualism of which they *prate*” (Roosevelt 1936: 16); “Я не знаю, *хватит ли ума* у господ из германского официоза догадаться, что ‘закрыть’ на бумаге то или иное государство они конечно могут, но, если говорить серьезно, то ‘сие от них не зависит’” (Stalin 1936: 24).

16 See examples: “Die nationalsozialistische Führung Deutschlands hat in dieser kurzen Zeit *ein Wunder* vollbracht” (Hitler 1936: 35); “Let them be specific in their negative *attack*” (Roosevelt 1936a: 16); “Таким образом, полная *победа* социалистической системы во всех сферах жизни народного хозяйства является фактом” (Stalin 1936: 8). Placing the last example into a “negative” category in the *Table 2* may raise eyebrows since it is not easy to detect pejorative connotations in the meaning of the word “victory”. However, if looked at from a communicative standpoint, this is exactly the case. In fact, as long as the sender acts in relation to its counterpart competitively rather than cooperatively, communication suffers and interactional asymmetry increases: in order to “win” (or “lose”), one has to have an “enemy”.

Since the crucial difference between antagonistic and agonistic asymmetries appears to stem from the latter's translatability into symmetrical terms, one may expect their semantics to possess some degree of transparency sufficient for conflation between internal (sender) and external (recipient) interpretations. Ironically, one of the sources of this transparency is the strong emotional undercurrent of some agonistic asymmetries: whereas the meaning of '*faulty*' could hardly result in shared understanding due to the impenetrable fusion between the sender's opaque but dominant Self and his or her evaluation motives, the individual semantics of such words as '*disturbing*', rooted in the speaker's reflection upon his or her own feelings, could blend with its normative meaning and evoke in the recipient the illusion of his or her affective similarity with the speaker. An alternative strategy of reconciliation between the conflicting interactional perspectives would be the translation of the speaker's reflexively accessed self into allocentric (systemic) terms: although such words as '*pacifying*' or '*procrastination*' reveal first and foremost the highly subjective evaluations of impact and tempo made on uncertain grounds, the underlying concepts informing these judgments—"peace" and "right tempo"—are disguised as intersubjectively proven and universally valid social phenomena. Notwithstanding these scenarios, agonistic asymmetries do not generally get transformed into symmetries but rather serve as empty promises of such transformations: although the gap between actual downplaying and pretended fixing of conceptual disparities is noticeably smaller in this milder version of asymmetrical concepts than in antagonistic asymmetries, it is by no means absent or semantically irrelevant.

Seen in this light, the distinctive feature of the *strong agonistic asymmetry* is its *de jure* openness to semantic negotiation, suggested by the sender to the recipient in full knowledge of the fact that a handful of standard communicative means would prevent this suggestion from having any practical consequences. Thus in the *sender-as-recipient*-model, the sense of affinity on both ends of the communication channel is created either by effusive externalization of sharable emotional states ('*reassuring developments*')[17] or by flashing com-

17 See examples: "Es ist daher *bedauerlich*, daß die übrige Welt für das Wesen und die Größe dieser Aufgaben dank einer ebenso leichtfertigen wie geistlosen, ja unnötig gehässigen Behandlung dieser Probleme kein Ver-

monly available data as a purported ground for evaluations ('*better than any other government in the world*'),[18] or even by keeping the meanings of evaluative terms as open and gradable as possible to simplify acceptance and ease negotiation ('*rather important changes*').[19] At the same time, the absence of actual communicative mechanisms which would make Self-perceptions of internal states comparable and criteria for arbitrary evaluations explicit, is downplayed at all cost: notwithstanding low comparability of both individual Self-perceptions (see above) and unstructured *ad hoc* judgments, the recipient is lead to take his feeling of being reassured for being similar with the sender's and to identify his way of favoring one government over others with the one of his communication partner. Alternatively, the *sender-as-system*-model simulates the commonality of interpretative perspectives by styling situational evaluations as universal formulas of systemic well-being (synonymous with '*unity*', '*dynamics*', '*stability*', '*order*', '*appropriateness*' etc.).[20] Consequently, the negative variant of strong

---

ständnis besitzt" (Hitler 1936: 37); "Among the Nations of the great Western Hemisphere the policy of the good neighbor has *happily* prevailed" (Roosevelt 1936a: 8); "В результате пройденного пути борьбы и лишений *приятно* и *радостно* иметь свою Конституцию, трактующую о плодах наших побед" (Stalin 1936: 29).

18 See examples: "Es ist ja auch kein Zufall, daß das Wirken des jüdischen Elements im Augenblick, in dem es glaubt, gegen den Staat sich ergeben zu können, um dessen Führung an sich zu reißen, zunächst *die größten bisherigen Gemeinschaftsleistungen der Staaten* zu beseitigen versucht" (Hitler 1936: 58); "*At no time in the four and a half centuries* of modern civilization in the America *has there existed* […] *a greater spirit of mutual understanding*" (Roosevelt 1936a: 9); "*Никогда в мире еще не бывало таких* действительно свободных и действительно демократических *выборов*, никогда!" (Stalin 1937a: 239).

19 See examples: "Der deutsche Lebensmittelmarkt kann durch die nun einmal gegebene Begrenzung unseres Bodens *nur ganz unwesentlich* gebessert werden" (Hitler 1936: 49); "Even though the individual results may seem smaller, the total is *significant*" (Roosevelt 1936p: 608); "Все эти и подобные факторы привели к тому, что *изменился в корне* облик народов СССР" (Stalin 1936: 14).

20 See examples: "Welches hat durch unseren *Emporstieg* etwas verloren?" (Hitler 1936: 35); "It has been our aim first of all to restore values to a

agonistic asymmetries ascribes the denial or subversion of these "unquestionably positive beliefs" (van Dijk 1988: 121) to the recipient (*'haste'*/*'arrested development'*/*'shaky foundations'*),[21] whereas their neutral version, devoid of ethical, deontic or epistemic connotations, merely indicates the distance between the speaker's understanding of the norm and the fact observed (*'small'*/*'quickly'*).[22] In contrast to antagonistic asymmetries which ban dissent outright, the strong agonistic asymmetries sweep it under the carpet: the common ground between communicators extends to sharable meanings—the lowest common denominator between two interpretations of a certain statement in a given speech act. Roughly speaking, as long as both communication partners find—for whatever reasons—the same things *'laughable'*, the same developments *'bad enough'* and the same tempo *'appropriate'*, the interaction may keep running more or less smoothly regardless of interpretational disparities. Regardless of this illusion of shared perception and evaluation,[23] the monopoly on evaluation rests with the

---

normal [1] and *proper* [2] level" (Roosevelt 1936i: 556); "Она не просто провозглашает право на труд, но и обеспечивает его законодательным закреплением факта *отсутствия кризисов* в советском обществе" (Stalin 1936: 14).

21 See examples: "Welche Rechtfertigung erhält unser ganzes Handeln gerade heute bei einem Blick in die *zerrissene* [1] und *haltlos gewordene* [2] Umwelt!" (Hitler 1936: 30); "The second thing we did was to help our *stalled* economic engine to get under way again" (Roosevelt 1936k : 559); "Теперь, наверное, все поняли, что *чрезмерное* [1] увлечение хозяйственными компаниями и хозяйственными делами *ведет к тупику* [2]" (Stalin 1937a: 238).

22 See examples: "Und es muß dann aber weiter festgestellt werden, daß gerade die für den Angriff geeigneten und bestimmten Waffen von den Partnern des Friedensvertrages *in der außerordentlichsten Weise* weiterentwickelt, verbessert und vermehrt werden" (Hitler 1935: 20); "I have belived for a *great many* [1] years that the time has come in our civilizazion wihen a *great many* [2] of these chances should be eliminated from our lives" (Roosevelt 1936n: 574); "Начать хотя бы с того, что наша промышленность выросла за этот периодв гигантскую силу" (Stalin 1936: 5).

23 The pioneers of interactional political science described this illusion—in unmistakably Freudian terms—as the identification of "target" (follower)

sender; however, disguised as *monopoly on externalizing internal states and choosing evaluation criteria*, it does not stand in the way of continuous interaction and therefore appears less solid as in antagonistic asymmetries.

Being the least confrontational form of asymmetrical concepts, the *weak agonistic asymmetry* is barely distinguishable from the general flow of consensual semantics on the level of a single utterance: as the sender *pro forma* invites the recipient to join efforts in reducing semantic complexity, the evaluative component of egocentric meanings—one of the biggest obstacles on the road to intersubjective agreement—is tucked away between the contextually and denotatively implied evaluative ("asymmetrical") reference and its metonymical "symmetrical" substitute in the text. In such a case, speaking about the treatise which '*gives room to many interpretations*' or invoking '*fantastic conception*' is clearly intended to be understood as a disagreement in the context of public political discourse, but this would hardly be the case if these phrases were uttered in an academic discussion of literary texts.[24] Hence the weak agonistic asymmetries present the most convoluted form of monopoly on evaluation, which is camouflaged as a *monopoly on choosing evaluative contexts for intrinsically non-evaluative statements*. Expectedly, this intricate arrangement has been as a rare bird as its direct opposite—the strong antagonistic asymmetries—which makes its sub-classification devoid of any practical purpose.

* * *

The arguments for integrating *asymmetrische Gegenbegriffe* into dissipative system of social communication would have made little prac-

---

with the "source" (leader) which ultimately makes social power work (French/Raven 1959: 161-163).

24 See examples: "Diese neue Deutschland *kann daher nicht in Vergleich gebracht werden* mit dem Deutschland der Vergangenheit" (Hitler 1935: 9); "They steal the livery of great national constitutional ideals to serve discredited *special interests*" (Roosevelt 1936a: 14); "Если взять капиталистические страны, то там между депутатами и избрателями существуют *своеобразные*, я бы сказал, довольно странные *отношения*" (Stalin 1937b: 191).

tical sense if correlations between the small- and large-scale social asymmetries were not examined empirically. To that end, the quadripartite classification of asymmetrical concepts proposed above was applied to the public discourse in societies with bigger (Nazi Germany, Soviet Union) or smaller (USA during Franklin D. Roosevelt's presidency) large-scale communicational asymmetries, commonly referred to as "totalitarian" and "democratic" states.[25] Past experience, some intuitions and common-sense expectations, resulted in the following main hypothesis:

The frequency and strength of secondary (evaluative) asymmetries in the semantics of public discourse may correlate with the strength of communicative asymmetries on a higher (systemic) level expressing the instability (complexity) of a political system.

The samples chosen for analysis were limited (at this stage) to the "senders" discourse. They consisted of Adolf Hitler's, Joseph Stalin's and Franklin Roosevelt's speeches delivered in 1935-1937 (Hitler 1935; Hitler 1936; Roosevelt 1936a; Roosevelt 1936b; Roosevelt 1936c; Roosevelt 1936 d; Roosevelt 1936e; Roosevelt 1936f; Roosevelt 1936g; Roosevelt 1936h; Roosevelt 1936i; Roosevelt 1936j; Roosevelt 1936k; Roosevelt 1936l; Roosevelt 1936m; Roosevelt 1936n; Roosevelt 1936o; Roosevelt 1936p; Stalin 1936; Stalin 1937a; Stalin 1937b).

The most general figures pertaining to the main hypothesis are presented in *Table 1.* It looks like the *absolute* number of asymmetries (line 5) does not correlate with any large-scale social asymmetries: it appears to be the smallest in Stalin' texts and the largest in Hitler's texts, with Roosevelt's data squarely in between. Nevertheless, it may be wrong to dismiss this information as mere reflection of general linguistic differences or individual stylistic priorities, since its correlation with the discursive reflections of leadership style is apparent. Indeed,

25 To be sure, the social asymmetries in the U.S. during FDR's presidency, being very moderate in comparison to Soviet Union and Nazi Germany, did overstretch the democratic norms (see Howell 2003; Schivelbusch 2005), but American political system proved capable of balancing its disequilibria by introducing a two-term presidency limit after Roosevelt's death.

Hitler who employs asymmetrical concepts in his speech more frequently than Stalin or Roosevelt, also has in his political discourse the biggest number of individual pronominal self-references (*'I'*-*'me'*-*'my'*) and the highest frequency of self-references as a chief executive (*'Führer'*/*'Kanzler der Nation'*/*'Chef der Reichsregierung'* etc.), with Roosevelt and Stalin trailing behind—in the same order as they do in *Table 1* (Postoutenko 2010b: 46-48). Although this parallelism does not correspond to those higher-level communicative asymmetries which sustain the very division between the "totalitarian" and "democratic" societies, such as low changeability of social roles, limited topicality of turn-taking in public communication and impossibility of repairs, (Postoutenko 2010a: 17-33), it does offer additional arguments for contrasting Stalin's impersonal ("bureaucratic") and Hitler's authoritarian ("traditionalist") ways of stabilizing charismatic leadership (see Kofler 1970: 39; Getty 1992; Postoutenko 2010b: 39-42; the terms in quotation marks are borrowed from Weber 1920: 143). It also raises the difficult question of Stalin's "invisibility" in his own discourse as a backdrop for his nearly absolutist rule (see the most recent examination: Khlevniuk 2010), which is partially addressed below in the discussion of the *Table 2*.

If the absolute number of secondary asymmetrical concepts does little to confirm or disprove the main hypothesis, the same cannot be said about their percentage breakdown by categories. Indeed, with the two extremes (the *strong antagonistic* and the *weak agonistic* asymmetries) almost extinct, the *weak antagonistic* asymmetries are noticeably more frequent in Stalin's and Hitler's speeches than in Roosevelt's texts. (Naturally, the distribution of the *strong agonistic* asymmetries is inverse). This reads like a tentative confirmation of the main hypothesis, which may be now formulated with better precision:

The frequency of secondary (evaluative) asymmetries in the public discourse goes hand in hand with the incidence of communicative asymmetries on a higher (systemic) level. More specifically, in so-called "totalitarian" societies such as Soviet Union and Nazi Germany, *both* the asymmetrical concepts granting the sender monopoly on evaluation (in ethical, deontic, epistemic, communicative and unspecified a priori terms) *and* the large-scale communicative asymmetries (restraining exchange of social roles, turn-taking and repairs) appear more often than in so-called "democratic" societies such as United States of America.

(One could comment in passing on the scarcity of strong antagonistic asymmetries: in all likelihood, even in the most oppressive systems the costs of sharing codes (verbal language) and practices (turn-taking) with the fiercest enemies were considered tolerable compared to the danger of having "non-humans" round the corner).

Since *Table 1* makes likely the special role of evaluative monopoly in discursive production and maintenance of "totalitarian" (as opposed to "democratic") society, the next logical step would be to inquire what sort of evaluation (*i.e* formulated in ethical, deontic, epistemic, communicative and unspecified *a priori* terms) was more likely to be monopolized in discourse of one or another leader, and if the micro-macro correlation of the kind established in the *Table 1* was also discernible at the level of single subcategories. But the data on *weak antagonistic asymmetries* presented in the *Table 2* suggests the possibility of such correlation in one case only: in both Stalin and Hitler, the monopoly of communicative evaluation is based on *others' flaws* (*negative* asymmetries) rather than *own virtues* (*positive* asymmetries). The absolute figures appear too small to pursue this case further.

Overall in the *Table 2*, the breakdown of asymmetries into subcategories appears to be uniform among the speakers, with only two deviations large enough to justify interpretation. Firstly, the number of epistemic asymmetries in general and *positive* epistemic asymmetries in particular in Stalin's discourse certainly stands out: in fact, almost a fifth of all weak antagonistic asymmetries and almost a tenth of *all* asymmetries in Stalin's speeches analyzed are self-referential professions of truth of the sender's discourse. It is hard to avoid the impression that in Stalin's discourse the often professed rationalism and scientism of Marxist doctrine prevails over its normative messianic zeal (see Widmer 1987: 74). Secondly, the number of *a priori* asymmetrical evaluations in Hitler's discourse is significantly higher than corresponding numbers for Stalin's and Roosevelt's speeches, which generally fits into the image of individual rule unrestrained by rational or moral considerations cultivated by Hitler from time to time with the help of his image-maker Goebbels (Kershaw 1987: 13).

The data on the *strong antagonistic asymmetries* in public speeches of Stalin, Hitler and Roosevelt, gathered in *Table 3*, complicates the balance of differences and similarities between speakers and social systems established in the previous table, without redrawing it entirely. The look at the asymmetrical presentations of *sender as recipient* makes it clear that the most basic asymmetry expressible by communicative means, described earlier as "the privileged access to own states and beliefs", is used sparingly both in the "totalitarian" and in the "democratic" public discourse. Apparently, extrapolating immediate self-perceptions (rather than more specific and impersonal evaluations) pays off only as long as such individual emotionality of the leader is considered to be valuable information in the social system in general. In this sense, Hitler's comparatively frequent recourse to this variation of strong agonistic asymmetry is not really surprising, being consistent with his unusually persistent possessive references to own body and emotions (*'my brain'*/*'my sensations'*/*'my patience'* etc.) (Postoutenko 2010b: 41). Some of the similarities observable in *Table 3* happen to be more enlightening than the expected differences. It seems remarkable, for instance, that in all three presentations of *sender as system*, *stability* ('Self-preservation') comes up approximately three times more often than *dynamics* ('pro-activity'), notwithstanding not only the anticipated gaps between "totalitarian" and "democratic" discourses, but also the already proved noticeable difference in Stalin and Hitler's leadership styles (see Weiss 2003: 319-320; Rolf 2004; Postoutenko 2010b).

Perhaps the last (but certainly not the least important) question that *Tables 2* and *3* may shed the light on is the overall modus of integration in the interwar societies seen through the prism of asymmetrical concepts: was this integration positive (centered upon the *goodness of the Self*) or negative (focused on *badness of the Other*)? For a while, sociologists stood by the first scenario, aided by linguists who gave positive terms better chances in language (Boucher/Osgood 1969), but today both of these optimisms seem to fade (Marczewska/Zagrodzki/Kurcz 1986; Nöth 1992; Luhmann 1995: 39-47). In this respect, the data on weak antagonistic asymmetries (*Table 2*) is not particularly informative, as the preferences given to the negative asymmetrical terms in the speeches of Hitler and Stalin are as minimal as the opposite tendencies detectable in Roosevelt's texts. However, these barely noticeable distinctions take a much more definite shape in *Table 3*: whereas positive and negative terms break even in the "totalitarian" discourse, the prevalence of the first over the second is unmistakable in Roosevelt's speeches (standing for "democratic" discourse). It might be tempting to conclude that "totalitarian" social systems favor the balance of positive and negative integration and "democratic" systems stand in this respect on the rosy side of life, but even if it's true, such generalization would deserve a separate discussion and a separate empirical study.

Table 1: General distribution of conceptual asymmetries in the public discourse of Soviet Inion, Nazi Germany and USA

	Soviet Union *Stalin speeches 1936-1937*	Nazi Germany *Hitler speeches 1935-1937*	USA *Roosevelt speeches 1936*
1. antagonistic strong (% to 1-4)	9 (1.5%)	16 (1.6%)	1 (0.1%)
2. antagonistic weak (% to 1-4)	258 (43.6%)	382 (37.8%)	202 (26.5%)
3. agonistic strong (% to 1-4)	321 (54.2%)	585 (57.8%)	551 (72.1%)
4. agonistic weak (% to 1-4)	4 (0.7%)	28 (2.8%)	10 (1.3%)
1-4 (% to the number of sentences)	592 (56.1%)	1011 (98.9%)	764 (76.7%)

Table 2: Weak antagonistic asymmetries: the detailed breakdown

	Soviet Union *Stalin speeches 1936-1937*	Nazi Germany *Hitler speeches 1935-1937*	USA *Roosevelt speeches 1936*
1. ethical positive *'right'* (% to 1-2)	8 (25.8%)	11 (55.0%)	23 (63.9%)
2. ethical negative *'crime'* (% to 1-2)	23 (74.2%)	9 (45.0%)	13 (36.1%)
1+2. ethical overall (% to 1-10)	31 (12.0%)	20 (5.2%)	36 (17.8%)
3. deontic positive *'necessary'* (% to 3-4)	14 (40.0%)	15 (77.3%)	13 (76.5%)
4. deontic negative *'inacceptable'* (% to 3-4)	21 (60.0%)	5 (22.7%)	4 (23.5%)
3+4. deontic overall (% to 1-10)	35 (13.6%)	20 (5.2%)	17 (8.4%)
5. epistemic positive *'truth'* (% to 5-6)	50 (79.4%)	44 (60.3%)	22 (61.1%)
6. epistemic negative *'false'* (% to 5-6)	13 (20.6%)	29 (39.7%)	14 (38.9%)
5+6. epistemic overall (% to 1-10)	63 (24.4%)	73 (19.1%)	36 (17.8%)
7. communicative positive *'clarity'* (% to 7-8)	8 (29.6%)	13 (23.2%)	16 (53.3%)
8. communicative negative *'clamor '* (% to 7-8)	19 (70.4%)	43 (76.8%)	14 (46.7%)
7-8. communicative overall (% to 1-10)	27 (10.5%)	56 (14.7%)	30 (14.9%)
9. *a priori* non-negative *'b(igg)est ever'* (% to 9-10)	40 (39.2%)	107 (50.2%)	35 (42.2%)
10. *a priori* negative *'enemies'* (% to 9-10)	62 (60.8%)	106 (49.8%)	48 (57.8%)
9+10. *a priori* overall (% to 1-10)	102 (39.5%)	213 (55.8%)	83 (41.1%)
(1+3+5+7+9). NON-NEGATIVE TOTAL (% to 1-10)	120 (46.5%)	190 (49.7%)	109 (54.0%)
(2+4+6+8+10). NEGATIVE TOTAL (% to 1-10)	138 (53.5%)	192 (50.3%)	93 (46.0%)

Table 3: Strong agonistic asymmetries: the detailed breakdown

	Soviet Union *Stalin speeches 1936-1937*	Nazi Germany *Hitler speeches 1935-1937*	USA *Roosevelt speeches 1936*
1a. sender-as-recipient: *extrapolated (Self-) perceptions ‘unfortunately’/‘laughable’*	22 (34.4%)	75 (75.8%)	34 (32.1%)
1b. sender-as-recipient *arbitrary(Other-)evaluations ’great’/‘mediocrity’* (% to 1)	42 (65.6%)	24 (24.2%)	72 (67.9%)
1. speaker-based total (% to 1-2)	64 (20.4%)	99 (16.9%)	106 (19.2%)
2a. sender-as-system: *system as (Self-)preservation ‘consolidation’/‘falling apart’ (% to 2)**	21 (8.2%)	102(21.0%)	41 (9.2%)
2b. sender-as system: *system as (pro)activity ‘moving forward’/’obsolete’ (% to 2)**	6 (2.3%)	38 (7.8%)	15 (3.4%)
2. system-based total (% to 1-2)	257 (79.6%)	486 (83.0%)	445 (80.8%)
NEUTRAL *[‘astonishing’/ ‘quickly’]* TOTAL (% to 1-2)	61 (19.0%)	110 (18.8%)	76 (13.7%)
POSITIVE *[‘great’/ ‘stabilization’]* TOTAL (% to 1-2)	130 (40.5%)	206 (35.2%)	301 (54.6%)
NEGATIVE *[‘laughable’/ ‘backwardness’]* TOTAL (% to 1-2)	130 (40.5%)	269 (46.0%)	174 (31.6%)

---

*2a and 2b do not add up to 100% because *(Self-)preservation* and *(pro)activity* are just two semantic variants of system-based strong agonistic asymmetries out of many.

REFERENCES

Primary Sources

Hitler, A. 1935. *Rede des Führers und Reichskanzlers Adolf Hitler vor dem Reichstag am 21. Mai 1935*. Berlin: M. Müller.

Hitler, A. 1936. *Reden des Führers am Parteitag der Ehre 1936*. München: Zentralverlag der NSDAP.

Roosevelt, F. D. [1936a] 1938. Annual Message to the Congress (January 3rd, 1936). In *Complete Speeches and Addresses. Volume 5*, by F. D. Roosevelt, New York: Random House, 8-18. (hereafter: *CS*).

Roosevelt, F. D. [1936b] 1938. Address at Chautauqua, NY (August 14th, 1936). In *CS*, 285-292.

Roosevelt, F. D. [1936c] 1938. Radio Campaign Address to Dinners of Business Men Held Throughout the Nation (October 23rd, 1936). In *CS*, 534-537.

Roosevelt, F. D. [1936d] 1938. Address at the Dedication of the New Chemistry Building, Howard University (October 26, 1936). In *CS*, 537-539.

Roosevelt, F. D. [1936e] 1938. Informal Extemporaneous Remarks at Dedication of Brooklyn College (October 28th, 1936). In *CS*, 539-540.

Roosevelt, F. D. [1936f] 1938. Address on the Occasion of the Fiftieth Anniversary of the Statue of Liberty (October 28th, 1936). In *CS*, 540-544.

Roosevelt, F. D. [1936g] 1938. Address at Roosevelt Park (October 28th, 1936). In *CS*, 544-546.

Roosevelt, F. D. [1936h] 1938. Campaign Address on John Mitchell Day (October 29th, 1936). In *CS, 546-549.*

Roosevelt, F. D. [1936x] 1938. Campaign Address at Harrisburg, PA (October 29th, 1936).In *CS*, 550-553.

Roosevelt, F. D. [1936i] 1938. Campaign Address at Camden, NJ (October 29th, 1936).In *CS*, 554-557.

Roosevelt, F. D. [1936j] 1938. Campaign Address at Wilmington, DE (October 29th, 1936). In *CS*, 557-558.

Roosevelt, F. D. [1936k] 1938. Campaign Address at Brooklyn, NY (October 30th, 1936). In *CS*, 559-563.

Roosevelt, F. D. [1936l] 1938. Informal Extemporaneous Remarks at a Gathering of Workers at National Campaign Headquarters (October 31st, 1936). In *CS*, 563/565.

Roosevelt, F. D. [1936m] 1938. Campaign address at Madison Square Garden, New York City (October 31st, 1936). In *CS*, 567-573.

Roosevelt, F.D. [1936n] 1938. Informal Extemporaneous Remarks at Poughkeepsie, NY (November 2nd, 1936). In *CS*, 573-577.

Roosevelt, F. D. [1936o] 1938. Final Campaign Radio Speech of the 1926 Presidential Campaign (November 2nd, 1936). In *CS*, 577-581.

Roosevelt, F. D. [1936p] 1938. Address before the Inter-American Conference for the Maintenance of Peace (December 1st, 1936). In *CS*, 604-611.

Stalin, J. 1936. [1936] 1945. *О проекте конституции Союза СССР. Доклад на чрезвычайном VIII Всесоюзном Съезде Советов* [November 25th, 1936]. Москва: Государственное издательство политической литературы.

Stalin, J. 1937a. Речь на предвыборном собрании избирателей Сталинского избирательного округа города Москвы ([December 11th, 1937]). In И. В. Сталин. *Сочинения*. Т. 14, Москва: Писатель, 238-243.

Stalin, J. 1937b. Заключительное слово на пленуме Центрального комитета ВКП (б) ([March 5th, 1937]). In И. В. Сталин. *Сочинения*. 14, Москва: Писатель, 189-207.

Secondary Sources

Berding, H. 1976. Begriffsgeschichte und Sozialgeschichte. *Historische Zeitschrift* 223: 98-110.

Busse, D. 1987. *Historische Semantik. Analyse eines Programms*. Stuttgart: Klett-Cotta.

Boucher, J./Osgood, Ch. 1969. The Pollyanna Hypothesis. *Journal of Verbal and Learning Behavior* 8 (1): 1-8.

Butterworth, G. 1995. An Ecological Perspective on the Origins of Self. In *The Body and the Self*, edited by J. L. Bermúdes/A. J. Marcel/N. Eilan. Cambridge, MA: MIT Press, 87-104.

Carlson, G. 2003. Interpretative Asymmetries in Major Phrases. In *Asymmetry in Grammar. Vol. 1 (Syntax and Semantics)*, edited by A. M. Di Sciullo. Amsterdam: John Benjamins, 301-314.

Chilton, P. 2004. *Analyzing Political Discourse*. London: Routledge.

Churchland, P. S. 2003. Self-Representation in Nervous Systems. *Annals of the New York Academy of Sciences* 1001: 31-38.

Davidson, D. 2001. *Subjective, Intersubjective, Objective*. Oxford: Oxford University Press.

Dijk, T. van. 1988. *Ideology. A Multidisciplinary Approach*. London: Sage.

French J./Raven, B. H. 1959. *Studies of Social Power*. Ann Arbor, MI: Institute of Social Research.

Gallie, W. B. 1956. Essentially Contested Concepts. *Proceedings of the Aristotelian Society* 56: 167-198.

Getty J. Arch. 1992. Les bureaucrats bolcheviques et l'État stalinien. *Revue des Études Slaves* 64 (1): 27-51.

Gloy, K. 1996. Recht haben. Ein Umweg zu einer Linguistik des Streitens. In *Öffentlicher Sprachgebrauch. Georg Stötzel zum 60. Geburtstag gewidmet*, edited by K. Böke/M. Jung/M. Wengerer. Opladen: Westdeutscher Verlag, 359-377.

Goodman, N. 1968. *The Languages of Art*. Indianapolis: Hacket Publishing.

Günther, H. 1979. Auf der Suche nach der Theorie der Begriffsgeschichte. In *Historische Semantik und Begriffsgeschichte*, edited by R. Koselleck, Stuttgart: Klett-Cotta, 102-120.

Guilhaumou, J. 2000. De l'histoire des concepts à l'histoire conceptuel des usages conceptuels. *Genèses* 38: 205-218.

Gumbrecht, H.-U. 2006. *Dimensionen und Grenzen der Begriffsgeschichte*. München: Wilhelm Fink Verlag.

Habermas, J. [1974] 1984. Überlegungen zur Kommunikationspathologie. In J. Habermas. *Vorstudien und Ergänzungen zur Theorie des kommunikativen Handelns*. Frankfurt am Main: Suhrkamp, 226-270.

Hass, U. 1991. Das Besetzen von Begriffen. Kommunikative Strategien und Gegenstrategien in der Umweltdiskussion. In *Begriffe besetzen. Studien des Sprachgebrauchs in der Politik*, edited by F. Liedtke/M. Wengerer/K. Böke. Opladen: Westdeutscher Verlag, 330-337.

Hobbes, Th. [1647]. 1966. De Cive. In *Opera philosophica quæ latine scripsit omnia*, by Th. Hobbes. Darmstadt: Scientia Verlag Aalen, 145-432.

Horwitch, P. 1987. *Asymmetries in Time*. Cambridge, MA: MIT Press.

Howell, W. G. 2003. *Power Without Persuasion. The Politics of Direct Presidential Action*. Princeton: Princeton University Press.

Ivanov, V. V. 1974. On Antisymmetrical and Asymmetrical Relations in Natural Languages and Other Semiotic Systems. *Linguistics* 119: 35-40.

Ivanov, V. V. 1983. *Gerade und Ungerade. Die Asymmetrie des Gehirns und der Zeichensysteme*. Stuttgart: Hirzel.

Jacob, F. 1982. *The Possible and the Actual*. Seattle & London: University of Washington Press.

Kershaw, I. 1987. *The "Hitler Myth". Image and Reality in the Third Reich*. Oxford: Oxford University Press.

Khlevniuk, O. V. 2010. *Хозяин. Сталин и утверждение сталинской диктатуры*. Москва: Росспэн.

Knobloch, C. 1992. Überlegungen zur Theorie der Begriffsgeschichte aus Sprach- und Kommunikationswissenschaftlicher Sicht. *Archiv für Begriffsgeschichte* 35: 6-24.

Kofler, L. 1970. *Stalinismus und Bürokratie. Zwei Aufsätze*. Neuwied: Luchterhand.

Koselleck, R. [1975] 1989. Zur historisch-politischen Semantik asymmetrischer Gegenbegriffe. In *Vergangene Zukunft. Zur Semantik geschichtlicher Zeiten*, by R. Koselleck. Frankfurt am Main: Suhrkamp, 211-259.

Koselleck, R. [1994] 2010. Feindbegriffe. In *Begriffsgeschichten. Studien zur Semantik und Pragmatik der politischen und sozialen Sprache*, by R. Koselleck. Frankfurt am Main: Suhrkamp, 274-283.

Koselleck, R. 2003. *Zeitschichten. Studien zu Historik*. Frankfurt am Main: Suhrkamp.

Leach, E. 1976. *Culture and Communication*. Cambridge: Cambridge University Press.

Lehmann, E. 1999. *Asymmetrische Information und Werbung*. Wiesbaden: Deutscher Universitäts-Verlag.

Lübbe, H. 2003. Wortgebrauchpolitik. Zur Pragmatik der Wahl von Begriffsnamen. In *Herausforderungen der Begriffsgeschichte*, edited by C. Dutt. Heidelberg: Universitätsverlag Winter, 65-80.

Luhmann, N. 1987. Soziale Systeme. Grundriß einer allgemeinen Theorie. Frankfurt am Main: Suhrkamp.

Luhmann, N. 1995. Jenseits von Barbarei. In *Gesellschaftsstruktur und Semantik. Studien zur Wissenssoziologie der modernen Gesell-*

schaft. Bd. 4, by N. Luhmann. Frankfurt am Main: Suhrkamp, 138-150.

Lyotard, F. 1983. *Le différend*. Paris: Minuit.

Marczewska, H/Zagrodzki, M./Kurcz, I. 1986. Asymmetry of Dimensions in Language Knowledge. *Advances in Psychology* 39: 69-93.

Missfelder, J.-F. 2006. Die Gegenkraft und ihre Geschichte: Carl Schmitt, Reinhart Koselleck und der Bürgerkrieg. *Zeitschrift für Religions- und Geistesgeschichte* 58 (4): 310-336.

Montaigne, M. de. [1595] 1931. *Essais*. Paris: Garnier frères.

Mühlhäusler, P./Harré, R./Freyne, M. 1990. Pronouns and People. The Linguistic Construction of Social and Personal Identity. Oxford: Basil Blackwell.

Neville, A. C. 1976. *Animal Asymmetry*. London: Edward Arnold.

Nöth, W. 1992. Symmetries and Asymmetries Between Positive and Negative Emotion Words. In *Anglistentag 1991*, edited by W. G. Busse. Tübingen: Max Niemeyer, 72-80.

Palonen, K. 2002. Begriffsgeschichte und/als Politikwissenschaft. *Archiv für Begriffsgeschichte* 44: 221-234.

Olsen, N. 2004. "Af alle mine lærere har Schmitt været den vigtigste". Reinhart Koselleck's intellectuelle og personlige relationer til Carl Schmitt. *Historisk Tidsskrift* 104 (1): 30-60.

Pankakoski, T. 2010. Conflict, Context, Concreteness. Koselleck and Schmitt on Concepts. *Political Theory* 38 (6): 749-779.

Postoutenko, K. 2007. How Narration Shapes History. Self-Descriptive Narratives and Their Social Functions. In *Varietas in Concordia. Essays in Honor of Pekka Pesonen*, edited by B. Hellman/T. Huttunen/G. Obatnin. Helsinki: Department of Slavonic and Baltic Languages and Literatures, University of Helsinki, 96-123.

Postoutenko, K. 2009. Between "I" and "We". Studying the Grammar of Social Identity in Interwar Europe (1900–1950). *Journal of Language and Politics* 8 (2): 195-222.

Postoutenko, K. 2010a. Prolegomena to the Theory of Totalitarian Communication. In *Totalitarian Communication: Hierarchies, Codes and Messages*, edited by K. Postoutenko. Bielefeld: transcript Verlag, 11-40.

Postoutenko, K. 2010b. *Soviet Culture. Codes and Messages*. München: Otto Sagner Verlag.

Rappaport, R. 1999. *Ritual and Religion in Making of Humanity*. Cambridge: Cambridge University Press.

Rolf, M. 2004. Working Towards the Centre. Leader Cults and Spatial Politics in Pre-war Stalinism. In *The Leader Cult in Communist Dictatorships: Stalin and the Eastern Bloc*, edited by A. Balazs/J. C. Behrends/P. Jones/E. A. Rees. Houndmills: Palgrave Macmillan, 141-157.

Said, E. 1979. *Orientalism*. New York: Vintage Books.

Schivelbusch, W. 2005. *Entfernte Verwandschaft. Faschismus, Nationalsozialismus, The New Deal (1933-1939)*. München: Hanser.

Schmitt, C. [1932] 2002. *Der Begriff des Politischen*. Berlin: Duncker & Humblot.

Schultz, H. 1979. Begriffsgeschichte und Argumentationsgeschichte. In *Historische Semantik und Begriffsgeschichte*, edited by R. Koselleck. Stuttgart: Klett-Cotta, 43-74.

Sciullo, A. M. di. 2003. Introduction: Asymmetry in Grammar. In *Asymmetry in Grammar. Vol. 1 (Syntax and Semantics)*, edited by A. M. Di Sciullo. Amsterdam: John Benjamins, 1-10.

Shannon, C./Weaver, W. 1949. *The Mathematical Theory of Communication*. Urbana, Il: The University of Illinois Press.

Sommer, R. 1969. *Personal Space.Tthe Behavioral Basis of Design*, Englewood Cliffs, NJ: Prentice-Hall.

Tauber, A. 1994. *The Immune Self. Theory or Metaphor?* New York: Cambridge University Press.

Tugendhat, E. 1979. *Selbstbewusstsein und Selbstbestimmung. Sprachanalytische Interpretationen*. Frankfurt am Main: Suhrkamp.

Villas Boas Castello Branco, P. H. 2006. A sociologia dos conceitos e a historia dos conceitos: um diálogo entre Carl Schmitt e Reinhart Koselleck. *Sociedade e estado* 21 (1): 133-168.

Weber, M. [1920] 1986. *Wirtschaft und Gesellschaft*. Tübingen 1986: Paul Siebeck (J. C. B. Mohr).

Wiener, N. [1948] 1984. Time, Communication and Nervous System. In *Collected Works with Commentaries. Volume IV*, by N. Wiener. Cambridge: The MIT Press, 220-243.

Wiese, L. von. 1965. *Die Philosophie der persönlichen Fürworter*. Tübingen: J. C. B Mohr (Paul Siebeck).

Weiss, D. 2000. Alle vs. Einer. Zur Scheidung von *good guys* and *bad guys* in der sowjetischen Propagandasprache. In *Slavistische Linguistik 1999. Referate des XXV. Konstanzer Slavistischen Arbeitstreffens*, edited by W. Breu. München: Otto Sagner Verlag, 237-275.

Weiss, D. 2003. Stalinistischer und nationalsozialistischer Propagandadiskurs im Vergleich. Eine erste Annäherung. In *Slawistische Linguistik 2001. Referate des XXVII. Konstanzer Slavistischen Arbeitstreffens,* edited by G. Freidhof/H. Kuse. München: Otto Sagner Verlag, 309-357.

Widmer, J. 1987. Aspects langagiers des totalitarismes. In *Forces et faiblesses des Totalitarismes*, edited by E. Morin. Fribourg: éditions Universitaires, 73-84.

Wittgenstein, L. [1951] 1989. Bemerkungen über die Farben. In *Werkausgabe. Band 8*, by L. Wittgenstein. Frankfurt am Main: Suhrkamp, 11-112.

Zeleny, M. 1980. Autopoesis. A Paradigm Lost? In *Autopoiesis, Dissipative Structures, and Spontaneous Social Orders*, edited by M. Zeleny. Boulder, CO: Westview Press, 3-43.

Three Takes on the Counter-Revolutionary

Studying Asymmetrical Political Concepts in the People's Republic of China

JUHA A. VUORI

INTRODUCTION: IDENTIFICATION OF THREATS AND THE JUSTIFICATION OF VIOLENCE

This chapter is used to discuss the most drastic function of asymmetrical political concepts, namely the legitimisation of violence or the use of force. Various political eras in the People's Republic of China (中华人民共和国 *Zhōnghuá Rénmín Gònghéguó* hereafter referred to as PRC) are replete with discourses that define various inner enemies for singularisation and subsequently for removal or even elimination. Indeed, labels such as 修正主义分子 (a 'revisionist', *xiūzhèngzhǔyì fè nzǐ*) or 反革命分子 (a 'counter-revolutionary element', *fǎngémìng fè nzǐ*) have had drastic consequences for the bearers of these labels (cf. Koselleck 2004: 155-157). Here, I will examine three conceptual tools which have been used to study asymmetrical concepts in the Chinese context. Carl Schmitt's friend-enemy dyad has been effectively used for this purpose by Michael Dutton (2005), Barend ter Haar (1997; 2002) has used the 'demonological paradigm', and I myself have employed the theory of securitisation (Paltemaa/Vuori 2006; Vuori 2008; 2010) to examine the political function of security arguments in China. I examine the benefits and shortfalls of such approaches within the empirical context of the Great Proletarian Cultural Revolution (文化大革命, *wénhuà dàgémìng;* CR). This should enlighten us on the role and political function of the 'counter-revolutionary' during this tumultuous political process.

The three approaches to the study of asymmetric concepts in politics examined here share similarities in that they serve the function of political singularisation. In processes of singularisation, exclusive claims at generality are made through self-definition that includes identity imputations and frames. The latter, in turn, affect counter-concepts that discriminate against the 'excluded' (Koselleck 2004: 156) imbued with negative identities. In Wittgenstein's (2001: §67) terms, such practices share 'family resemblances'. The differences of these three familiar approaches affect how applicable they are for the comparative study of societies.

The type of singularisation discussed here functions to identify threats.[1] This kind of identification is a significant means to legitimise the use of force: drastic measures and transgression of 'rules' may be deemed as allowed in order to repel threats. Such means of protection against threats are often called security. However, the study of how issues become matters of security has uncovered that in order for an issue to be 'securitised' or for it to follow the logic of security, the word security itself does not have to be used. Certain words or concepts (e.g., terrorism) automatically bring the logic of danger, vulnerability, and fear with them which means that the necessity to combat them does not need to be elaborately argued each and every time. The use of such watchwords, or "institutionalised securitisation", as Buzan/Wæver/de Wilde (1998: 27-29) call the phenomenon, reduces the need for elaborate arguments on the securityness of specific cases. Indeed, the continuous use of watchwords (like 'counter-revolution',

1 Anne Thurston (1990) has analysed the use of violence during the Cultural Revolution on three levels: 1) a permissive level that contains the long-term, underlying factors that make the use of violence possible, 2) the proximate level, i.e., those social and ideological conditions that foster and legitimise violence, and 3) the immediate conditions that lead individuals to act violently. The use of asymmetric political concepts falls into the second category here.

'socialism', or 'terrorism') can be seen as an indicator of a successfully institutionalised securitisation.[2]

Politicians can proclaim to maintain 'security', which is desirous if compared to insecurity or outright chaos. Murray Edelman (1972: 9) has noted that governments which force unwelcome changes in their subjects' behaviour have the greatest need for reassuring symbols. Security as 'stability and unity' is especially potent in China where 'chaos' or 'turmoil' has been a recurrent fear throughout different eras of politics (Pye 1992: 12-16; Buzan/Wæver 2003: 140, 152). In China, stability and unity are not merely slogans of leadership; they are embedded in the Chinese 'collective memory' through many forms of culture and tradition.[3] The fear of chaos has often led to an overwhelming emphasis on ideological consensus in China. Indeed, the restoration of harmony (和谐, *héxié*) was of major importance in Confucian philosophy.

In imperial China, many rituals and doctrines that dissented from the Confucian cosmological order became targets of government suppression; religious sectarian groups that espoused beliefs and practiced rituals which the authorities deemed heterodox were a major governmental concern (Shek 1990: 87). Accordingly, China has had an im-

---

2 This was especially prevalent during the Cold War, as for example noted by Murray Edelman (1972 [1964]: 15): "'USSR' and 'Khrushchev' can come to stand so repeatedly for danger that adaptive thinking becomes unlikely, and political actions that accept the USSR or Khrushchev as reasonable or as potential associates are met with hostility". While this passage was about the U.S. in the 1960s, the same could be said about the PRC of the 1960s as well.

3 The patriotic education campaign after 1989 has, for example, presented the Communist Party of China (中国共产党, *Zhōngguó gòngchǎndǎng*, hereafter referred to as CCP)as a saviour of the nation in various course materials in the form of stories of heroes of the revolution and the struggle against Japan (see e.g., Hughes 2006: 73-76; Wang 2008). These stories present a narrative of the CCP as the bringer of the Motherland to light from the century of humiliation by foreign powers. When China has been unified, it has been strong, while when China has been in disarray, it has been dominated by foreign powers. This is meant to infuse students with a strong sense of 'stability and unity' under the leadership of the CCP as equated to patriotism and love of the Motherland.

pressive number of quasi-religious popular uprisings: in the 18th and 19th centuries almost every popular uprising was in some way related to religious movements (Yang 1961). In addition to unrest and civil strife, many religiously justified uprisings actually questioned the cosmological order on which the imperial system was based and thereby the whole political system of rule.

Barend ter Haar (2002) argues that the CCP has retained the practice of such a division of people into categories of 'good' and 'evil'. These practices have been used to legitimise the use of violence, which has been a regular aspect of Chinese social life and politics, even before the 20th Century. This violent aspect of Chinese society is dismissed in much cultural literature on China that emphasises the harmonious elements of Chinese philosophy and religions. According to him, the Yin-Yang principle[4] is used to hide the fact of violent conflict in China, the fact that more often than not harmony is not achieved, and when it is, harmony is a result of the subjugation of the many by the few. The use of asymmetric political concepts has indeed been prevalent in the PRC.

The chapter progresses through three takes on how such asymmetric labels that have been employed to legitimise use of violence in the PRC operate. These discussions are followed by a chapter on the tactics of political singularisation that can be discerned from the Cultural Revolution.

TAKE ONE: THE STATE AS EXCEPTION AND THE POLITICAL IN CHINA

When politics and the legitimisation of the use of force are discussed, Carl Schmitt's notions of the political and the state of exception often pop up. Schmitt (e.g., 1996; 2005) discussed the state of exception in the context of constitutional legal systems. For him, the essence of the political is the moment when normative procedural constraints on political power have to give way to the necessity of dealing with the en-

4 Or the 太极 (*tàijí*), the great or ultimate, the origin of the 阴 (*yīn*) and the 阳 (*yáng*).

emy.[5] This reveals the primacy of the exception over the norm.[6] The power to decide on such an exception from regular procedures and limits defines a sovereign[7] and marks the authentically political moment of creation.[8] For Schmitt, the authentically political moment of creation boils down to a struggle between (Christian) good and evil; for Schmitt (2005: 36): "all significant concepts of the modern theory of the state are secularized theological concepts"; "every political idea in one way or another takes a position on the 'nature' of man and presupposes that he is either 'by nature good' or 'by nature evil'" (Schmitt 2005: 56). This is evident in his assumption of the existence of basic dyadic relations: "Let us assume that in the realm of morality the final distinctions are between good and evil, in aesthetics beautiful and ugly, in economics profitable and unprofitable" (Schmitt 1996: 26). In the realm of politics, then, the basic dyad is the famous 'friend' vs. 'enemy'. With the help of this vitalist philosophy, Schmitt sought the authentic political condition or experience, which was the definition of the enemy and the decision on the state of exception—the real possibility of 'physical killing' (Schmitt 1996: 33; Huysmans 2006: 135-138).[9] Hence Schmitt's idea of the political was about the most drastic of exclusive relations—those of mortal enemies.

Michael Dutton (2005) has applied Schmitt's concept of the political to study Chinese policing. He demonstrates that campaigns against various enemies were so frequent in Mao's China that they might be understood as a formative element of the system rather than as exceptions (cf. Agamben 1998; 2005). Indeed, the cardinal element of Mao-

5 Schmitt (1996: 26): "The specific political distinction to which political actions and motives can be reduced is that between friend and enemy."

6 Schmitt (2005: 5): "This definition of sovereignty must therefore be associated with a borderline case and not with routine."

7 "Sovereign is he who decides on the exception" (Schmitt 2005: 5). The sentence in the original German is "*Soverän ist, wer über den Ausnahmezustand entscheidet*", which implies the possibility of the sovereign in plural.

8 For Giorgio Agamben, the political is the nexus between violence and law (Agamben 2005: 88), and "in modern biopolitics, sovereign is he who decides on the value or nonvalue of life as such" (Agamben 1998: 142).

9 Schmitt (1996: 29): "The political is the most intense end extreme antagonism."

ist politics was the almost continuous series of campaigns against various types of enemies. This is already evident in the opening sentence of the Selected Works of Mao Zedong: "Who are our enemies? Who are our friends? This is a question of the first importance for the revolution."[10] Dutton (2005: 3) uses this as the platform from which to launch his analysis of the tradition of policing in China. Using Schmitt's principle of enemy definition as the essence of the political, Dutton argues that this was indeed the essence of policing in China before the establishment of the PRC, which continued to persist under Mao's rule and ended only with his demise.[11]

Here, it must be enquired as to what exceptionality means in political orders that are not liberal democratic, or more precisely, what an exception means in a non-democracy where the legal order is not settled but remains in continuous flux.[12] For example, why do non-democratic orders declare martial law when they could use their prerogative power anyway? Mark Neocleous (2007: 14) offers an answer: governments seem to prefer to operate under the cloak of legality, so that, in alternative to suspension of law, violent actions of state apparatuses conducted under 'emergency conditions' have been legalized on the grounds of necessity and in the name of security. Indeed, it would seem that even in political orders in which the state is ideologically legitimised as an exception, the adherence to laws, decrees, and other norms remains necessary for legitimizing the use of force just

10 谁是我们的敌人?谁是我们的朋友?这个问题是革命的首要问题.

11 Indeed, Mao's practice to define and search for enemies seems to concur with Schmitt's friend/enemy dyad. Yet, as MacFarquhar/Schoenhals (2008: 201) observe, the centrality of the friend/enemy distinction for the CCP may be actually emanating from Lenin's writings—more specifically, from his Kossack dictum Кто кого? or 'Who will prevail over whom?' Lenin's dictum was used in China, for example, by Liu Shaoqi in his discussions with Mikojan in the 1940s (Heinzig 2004: 149), and was also employed in the Polemics (1965: 471-472) between the CCP and the Communist Party of the Soviet Union: "[A] very long period of time is needed to decide 'who will win' in the struggle between socialism and capitalism". See Palmujoki (1995: 39-40) on how the dictum was adopted (via Stalin) and operated in North Vietnam's political language and practice.

12 Žižek (2002) has noted that totalitarianism is mired in a liberal-democratic notion of normalcy.

once the initial stage of the revolution has been accomplished. After this point, the use of force may in fact be read as a sign of weakness.[13]

Lenin (2004: 13-18) famously argued for withering away of the state, done in three steps:[14]

> "The first stage—the bourgeois state: in capitalist society, it is the bourgeoisie that needs the state. The second stage—the state of the dictatorship of the proletariat: during the period of transition from capitalism to communism, it is the proletariat that needs the state. The third stage—the communist society: the state is not necessary, it withers away."

This doctrine implies that the state as such is a 'state of exception'; thereby the dictatorship of the proletariat, as a form of state, is the state of exception that is used to engender the stateless mode of existence. For Agamben (2005: 2), modern totalitarianism is a form of legal civil war that allows for the elimination of not only political adversaries but whole categories of people; what for Marx was a class struggle, turns in modern totalitarian regimes into a constant war which can only end when the 'People' and the people are merged into a classless society or a messianic state (Agamben 2001: 31). The state of exception is the means to accomplish this. In these terms, the Chinese case presents the state operating in a state of exception: the state is the means to eliminate the state through the elimination of classes and thus achieve the

13 The Cultural Revolution, however, appears to go against such a viewpoint: Mao appears to have sought the continual retention of the *nomos* i.e., the pure immediacy of rule unmediated by law. Indeed, the Cultural Revolution effectively dissolved the rules that had been instituted within the CCP. Overall, Mao's preferences seem paradoxical: he seemed to relish upheaval (乱 *luàn*) and accordingly turned the party inside out while he criticised bureaucratism and 'ecclecticism', yet also appreciated the benefits of a proper bureaucracy that could provide him with information and service (MacFarquhar/Schoenhals 2008: 188).

14 The original idea of the state withering away is from Engels, but it seems that Lenin's explanation of what this means vis-à-vis the state and the revolution has been more influential. This quote is from the most important of the Polemics (1965: 448) between the Chinese and the Soviet Communist Parties.

'withering away' of the state itself, as classes and the civil war between them would cease to exist.

A problem with Dutton's the application of Schmitt's friend/enemy dyad to Chinese politics is that only one aspect of Schmitt's metaphysics is taken into account:[15] he fails to deal with the issue of exceptionalism and the issue of constitutive and constituted power. Dutton seems to implicitly identify a problem with Schmitt's metaphysics applied to the case studies of politics: Dutton's application of the dyad rests on a metaphysical foundation different from that of Schmitt. For Schmitt, to define the dyad is a decision, yet Dutton (e.g., 2005: 77-78) describes the dichotomy as a flexible discourse unit. For example, during the anti-Japanese war, the core of the enemy definition shifted from 'class' to the 'nation' (Dutton 2005: 77-78). It seems that for Dutton (2005: 80), the political is not a decision, but rather a "dyadic process of framing politics". This raises the question as to whether some other approach to that of Schmitt might be more suitable for analysing political functions of asymmetric concepts.

Dutton's study has uncovered an important feature of Mao's brand of politics. But it seems that the enemy/friend distinction is not enough for an explanation of the totality of Chinese politics during Mao's lifetime. Enemy definition is better viewed as a technique, a mechanism, or a logic, which may also have unintended consequences. Other means beyond the dyad are needed to provide greater nuance for the study of such techniques. For Dutton, the dyad takes centre stage and seems to explain everything in Chinese policing. I, on the contrary, argue that while the construction of impressions of threats and the legitimisation effect of claims to repel them is an important political technique, the effects of these kinds of tactics and strategies are not the only matter that affects politics or politicking. Indeed, as Dutton's use of Schmitt's notion of the political suggests, the political is too broad a situation to shed light on the more tactical and limited use of asymmetrical political concepts in the legitimisation of violence. Perhaps it might be preferable to examine the use of asymmetric political concepts through other means rather than essentialising the dyad as such.

15 Dutton (2005: 304): "I have no interest in Hitler's so-called crown jurist beyond this."

TAKE TWO: THE TRADITION OF THE DEMONOLOGICAL PARADIGM

Whether or not Mao's keen sense to define enemies was influenced by either Schmitt's or Lenin's similar ideas, Chinese practices of enemy definition and branding predate Mao Zedong. Barend ter Haar (1997; 2002) has discussed Chinese traditions of the creation of 'us' as opposed to the threatening 'other'. He suggested that Chinese politics before the declaration of the People's Republic operated under a 'demonological paradigm'. Furthermore, in his view, the explicit forms of this *modus operandi* (or at least its operational logic) have been transported from the imperial era into the PRC times. Haar argues that violence has been a regular aspect of Chinese social life and politics, even before the 20th Century, a factor which is dismissed in much cultural literature on China (which usually emphasises the harmonious elements of Chinese philosophy and religions). Finally, Haar reasons that the Yin-Yang principle is used to obscure the reality of violent conflicts in China. More often than not, harmony is not achieved, and even when it is, the price is usually the subjugation of many to the few.

The demonological paradigm has been used in China to form a dyad of an 'us' (天下, *tiānxià*, all under heaven) and a 'them' (妖怪, *yāoguài*, demons) which seems to have parallels with Schmitt's friend-enemy distinction. At times of political strife—for example, during the 19th Century Taiping rebellion—the targeted individuals and groups could be defined in demonic terms, after which they could be fought with anti-demonic means and become the target of the most severe violent acts (Haar 2002: 36, 47). For example, from 1850 onwards the Taipings referred to both external conflict and internal strife in demonic terms; the Taipings were the divine army needed to combat demons, while individuals that ranged from local opponents to the Manchus were represented as demons (Haar 2002: 47; see also Haar 1997).[16]

16 While Chinese sectarian scriptures often portray apocalyptic destruction as a cosmic event principally carried out by spirits and demons, some sectarians, like the Taipings, have felt compelled to take it upon themselves to commit violent acts in expeditions aimed at approaching the Millennium (Shek 1990: 103). Not all Chinese sectarian groups have rebelled or become violent, and not all millenarian religious groups have resorted to violence (Rinehart 2006: 26). Yet, the urgent and immediate eschatology of

This demonological paradigm may also explain Mao's use of quasi-religious language like 'ox-monsters,' 'snake-demons,' 'evil spirits,' and 'poisonous weeds' in his political rhetoric, and practices like 'loyalty dances' that were prevalent in Maoist politics (Perry/Selden 2003). Such concepts and practices[17] have long traditions and awareness of demonic folklore was widespread and provided cultural resonance for the guidance of popular traditions into support for the CCP. For example, the ox demon (牛鬼, *niugui*) which appears in many folk tales, was used to dehumanise people; in the CCP's usage it labelled its targets as threats to society as well as backward, in a time when there was a call to "smash the old and build the new" (Haar 2002: 56).

During the Cultural Revolution, the youth, organized into the Red Guards, was supposed to carry out the violent struggle against the dangers that threatened China (Thurston 1990). In order to make its national level-discourse more efficient, the CCP utilised in this struggle identity frames[18] which resonated at the local level. As Barend ter Haar (2002: 56-57) notes: "This discourse [the demonological paradigm] helped create the Other that could be dealt with aggressively outside the customary channels for solving political conflicts, which were perceived as too gradual". Indeed, many peasants used the opportunity to 'settle old scores' with local administrators after Mao's call to "bombard the headquarters".

The CCP has claimed to be against all forms of anti-scientific superstitions inclusive of many anti-demonic practices prevalent in Chinese folk-beliefs. Yet, the party has not been adverse to label some individuals or social groups as 'bad elements', to call for their removal

sectarian groups makes their subversive potential superior to that of the orthodoxy (Rinehart 2006: 108).

17 In China, public morality is connected to security on the level of individuals or smaller segments of society: the deceased were thought to be able to become vengeful ghosts or forces of protection, depending on the behaviour of individuals and groups in respect to the "proper" moral stand (Radtke 2008: 204).

18 Frames are understood here as the interpretive schemata that simplify and condense "the world out there by selectively punctuating and encoding objects, situations, events, experiences, and sequences of actions in one's past" (Snow/Benford 1992: 137).

from society, and even their elimination by drastically violent means (Haar 2002: 54).[19] As evidenced by the frequency of purges within the CCP and in Chinese society in general (see for example Dutton 2005), the practice to 'demonise' individuals has not vanished: 'demons' were replaced with 'struggle' and 'counter-revolution' (Haar 2002: 54). Indeed, during the Mao era, which frequently employed the grammatical forms that belong to the demonological paradigm, counter-revolution was a violent threat that had to be responded to and dealt with proportionate counter-violence. Even though the threat was different to that of the imperial era, violence remained *the* means to exorcise the 'evil'.

The actual referent of the threat that China faced was often left open, which allowed for its convenient modifications in various situations. Such obscurity concurs with political practices beyond China: intangible threats allow polarised role-taking, which, in its turn, permits exaggeration and the justification of unfavourable policy choices (Edelman 1972: 71). If the claimed threat is clearly observable, tangible and undergoes systematic scrutiny, perceptions of its character converge with the techniques required to deal with it. This would make the political use of threats more difficult, which may be a reason why intangible threats may be preferable to the definite, tangible ones.

The demonological paradigm provides us with insights into the tradition of using asymmetrical concepts in China, but by the same token it also remains specific and limited to this context. If we want to gain a broader grasp of how asymmetric concepts operate to legitimise the use of violence, we must be able to distance ourselves from specific contexts. Indeed, similar practices of legitimisation are evident in many other societies beyond China, which means that other conceptual approaches are necessary to provide the means for comparison.

19 Cf. Schmitt's (1996) concept of *hostis*: all systems retain the possibility to decide upon inner enemies.

TAKE THREE: SECURITISATION AND THE USE OF ASYMMETRIC POLITICAL CONCEPTS

Exclusionary identities and the delineation of antagonistic Others has been identified as a major feature of 'security speech' within Security Studies. One of the most elaborate approaches that has been developed to examine precisely such processes is the theory of securitisation (Wæver 1995; Buzan/Wæver/de Wilde 1998; Balzacq 2010). This theory or model describes the frame, script, plot or grammar of security—the way we have learned to understand what *security* is, what *is* security, and how something becomes security. The transformation of issues into matters of security is called securitisation, which is conceptualised as an illocutionary speech act. In such moves an existential threat is 'produced' in relation to a referent object; an act of securitisation is to classify an issue as an existential threat which requires drastic measures.[20] If securitisation is 'successful', legitimacy or some other perlocutionary effects sought by the enunciator are created by means of a wider social process. Such a process consists of increasing and possibly escalating the instances informing the acts of securitisation (securitisation moves), which enables the speaker to 'break the rules' that normally constrict behaviour and policies (Vuori 2008). This, in turn, allows the question to shift into an area of 'special politics'—the politics of utmost priority. The framework of securitisation directs our attention to how issues of security are 'constructed' in the social fields of practice. Such investigations begin with speech acts that constitute securitisation moves.

From the viewpoint of conceptual history, the analysis of sequences of speech acts that are deemed to constitute securitisation moves may allow the examination of continuity and change in how concepts are used to produce the most exclusionary of others. The study of

20 Buzan (2008: 553): "Securitization is when something is successfully constructed as an existential threat to a valued referent object, and that construction is then used to support exceptional measures in response"; Wæver (2008b: 582): "Securitization is the discursive and political process through which an intersubjective understanding is constructed within a political community to treat something as an existential threat to a valued referent object, and to enable a call for urgent and exceptional measures to deal with the threat."

speech acts and language in a more general sense are important tools to identify conceptual changes at certain moments, or over periods of time (Brauch 2008: 67; Wæver 2008a: 100). Here, securitisation can been viewed as a 'grammar' of a certain type of human interaction. 'Resonant values' in certain situations then form the 'vocabulary' to effectively produce matters of security (Paltemaa/Vuori 2006; Vuori 2010). In Mao's China, counter-revolution was an institutionalised basis for securitisation into which particular instances and chains of events could be grafted. This demonstrates how, in one way, social artefacts—issues of security, in this case, are built into the 'background' of social reality. Although labels like 'turmoil' and 'well planned plots' would not seem to fit well into European political rhetoric, the language Chinese officials have used to construct official security realities is remarkably consistent with the 'grammar' that the theory of securitisation would predict, which makes it unnecessary to adapt 'culturally alien' concepts to the theory.

When some of the most important documents of the initiation of the Cultural Revolution are examined by the means of this framework, a security argument becomes apparent. In the so called May 16th Circular (Zhongfa 267) Peng Zhen, a leadership figure in the party and a political opponent of Mao, was singled out, together with the other authors of the so called February Outline, as "*a bunch of counter-revolutionaries opposing the communist party and the people*".[21] Their activities were presented as severe threats for the party: "*their struggle against us is one of life and death, and there is no question of equality. Therefore, our struggle against them, too, can be nothing but a life-and-death struggle, and our relation to them can in no way be one of equality*".[22] An element of the future was also present: "*Far from being a minor issue, the struggle against this revisionist line is an issue of prime importance having a vital bearing on the destiny and future of our party and state, on the future complexion of our party and state, and on the world revolution*".[23] The Circular further warned that the

21 "他们是一群反共、反人民的反革命分子" (Zhongfa 267).

22 "他们同我们的斗争是你死我活的斗争,丝毫谈不到什么平等." (Zhongfa 267)

23 "同这条修正主义路线作斗争,绝对不是一件小事,而是关系我们党和国家的命运,关系我们党和国家的前途,关系我们党和国家将来的面貌,也是关系世界革命的一件头等大事" (Zhongfa 267).

bourgeoisie had infiltrated the party and planned to restore capitalism and turn the dictatorship of the proletariat into a dictatorship of the bourgeoisie: "*Those representatives of the bourgeoisie who have sneaked into the party, the government, the army, and various cultural circles, are a bunch of counter-revolutionary revisionists*".[24] Labels such as 'reactionaries', 'class enemies', 'bad elements', 'splitters', 'extremists', and 'imperialists' carry national security connotations with them (some domestically and others internationally).[25] Here, counter-revolution was synonymous with national security.

Lin Biao's[26] (1966) speech a few days later drove this security argument home. Lin claimed that Peng Zhen planned a counter-

24 "混进党里、政府里、军队里和各种文化界的资产阶级代表人物,使一批反革命的修正主义分子" (Zhongfa 267).

25 MacFarquhar/Schoenhals (2008: 304) note how during the Cultural Revolution, almost anything could be 'upgraded' to a more serious crime by addition of the attribute of 'underground' which implied a degree of organisation which often was not the case. Similarly, any undesired activity that involved more than one or two people could be labelled as a 'gang'. The use of counter-revolutionary threats outlasted Mao's period of rule. For example, Deng Xiaoping's (1995: 176-177) speech that defined the core ideological principles of post-Mao China by promulgating the Four Cardinal Principles (四項基本原則, *sì xiàng jīběn yuánzé*) identified the enemy that the 'dictatorship of the proletariat' was directed against, as an amalgam of "*counter-revolutionaries, enemy agents, criminals and other bad elements of all kinds who undermine socialist public order, as well as new exploiters who engage in corruption, embezzlement, speculation and profiteering*". The existence of such institutionalised enemies made the continued existence of the state and its systems of control legitimate: "*so long as class struggle exists and so long as imperialism and hegemonism exist, it is inconceivable that the dictatorial function of the state should wither away, that the standing army, public security organs, courts and prisons should wither away.* [...] *The fact of the matter is that socialism cannot be defended or built up without the dictatorship of the proletariat*".

26 During the Cultural Revolution, Lin Biao was Mao Zedong's chief acolyte in charge of the People's Liberation Army, Mao's 'close comrade in arms' and even named as the heir apparent in the party's constitution. However, Lin soon fell in disfavour with Mao, and eventually died under mysterious

revolutionary *coup d'état*; the greatest issue in the question of Peng Zhen and the others under discussion was the security of the revolution and the socialist state: "*the big issue is preventing a counter-revolutionary coup d'état and being on guard against subversion*".[27] Lin then requested that the members of the standing committee remain vigilant and keep their guard up. Such deeds would provide security: "*This will safeguard us against revisionism and prevent capitalist restoration. This is the utmost essence, the crux of the issue*".[28] In practice this meant being in opposition to Peng Zhen and the others and maintaining Mao Zedong Thought and Chairman Mao as the absolute ruler of the party and the state. The acceptance of such moves provided legitimacy for continued struggle, the purge of leading party members, and the supreme position of Chairman Mao.

Such speeches illustrate the distinct security argument which gave the impetus to the political processes normally called "the Cultural Revolution": it was deemed legitimate to remove large parts of the top leadership and (for the 'masses') to attack party authorities because counter-revolutionaries had 'sneaked' into the party. The threat of a coup was used as the basis to legitimise the 'Cultural Revolution' that 'broke the rules' of the constituted political order of the CCP and the PRC. The Cultural Revolution was not only about security, yet this security argument and the confrontational politics of the era have had their influence on the political struggles which have contained elements of regime, or political security.

The PRC has its own set of institutionalised master signifiers, or watchwords, of security. The logic of such institutionalised categories can remain constant, but the signifiers that refer to institutionalised signifieds are changeable. The fate of Lin Biao is the case in point: after he was alleged to have staged a coup attempt and tried to defect to the Soviet Union, 'Lin Biao' was transformed from originally a chief securitising actor and even Mao Zedong's heir apparent in the late 1960s, to its diametrical opposite—an institutionalised signifier of 'counter-revolution' in the early 1970s.

---

circumstances in an apparent defection attempt to the Soviet Union in 1970.

27 "...,是防止反革命政变,防止颠覆的大事" (Lin 1966).

28 "这样,就能保证我们防止修正主义,避免资本主义复辟。这是最最根本的关键问题" (Lin 1966).

THE TACTICS OF POLITICAL SINGULARISATION AND THE LEGITIMISATION OF RULE-BREAKING DURING THE CULTURAL REVOLUTION

The Cultural Revolution has been interpreted both as an ideological struggle within the party and as a means to purge and reshuffle its top leadership. While Mao's social and political authority was essential to launch and maintain the process, he was not its sole architect. The Leftist faction that came to dominate the Cultural Revolution Group tried to steer the movement in various directions, as did Lin Biao's military faction. While these groups used various tactics in their struggles to gain Mao's favour and thus to improve their standing in the party hierarchies, those are security arguments and the identity frames used in these power-plays that are of relevance here.

The party and the People's Republic as a whole had already witnessed several 'line struggles' and purges or 'rectifications'. To label someone a leader of an erroneous ideological line or even the head of a hidden 'headquarters' within the party was not unprecedented. Such tactics involved security logic: those unwilling to abide by the Leninist principles of centralist democracy constituted a threat to the party. In terms of securitisation theory, there was a set of 'master signifiers' that had been 'institutionally securitised' in the previous purges the party had been through. Names and labels given to ourselves and others matter a great deal, as they articulate people's relations to others (Koselleck 2004: 155-157). The use of such master signifiers is an example of how 'asymmetric' classifications have been used in Chinese politics. The concepts like 'erroneous line [of thought and/or action]' exclude certain groups from the core of the party and thus, concomitantly, creates a unity within it. This serves the purpose of 'political and social singularisation' (Koselleck 2004: 156).

A prime example of such tactics is the singling out of Peng Zhen. With the help of Jiang Qing[29], Mao apparently set an elaborate trap for

29 Jiang Qing was Mao's wife who rose to a major position of power outside the party establishment by means of the Cultural Revolution. Premier Hua Guofeng staged an abortive coup against her and her supporters in 1976 a

Peng, who remained unaware of it until too late.[30] Importantly, the use of the 'grammar of security' was the final stage in this tactic to purge Peng and the others (i.e., Luo Ruiqing, Lu Dingyi, Yang Shangkun and his wife);[31] to label them as security threats was a move that placed their fate unreservedly in the hands of Mao. Moreover, the tactic provided a simple legitimisation narrative for the general public: sudden changes in the consolidated party leadership need not be interpreted as a sign of factional power-plays, but rather as a necessity for national security and the survival of the revolution. The unity of the party would be maintained, and to single out of forces within, but in league against it, served precisely this function.

few months after Mao's death. She was convicted of counter-revolutionary crimes in 1980 and died in prison in 1991.

30 This trap was set by the publication of a critique of Wu Han's play *Hai Rui Dismissed from Office* without the authorisation of the Propaganda Department. Any of Peng's possible actions, be it neglect or endorsement of the article, was bound to get him into trouble (MacFarquhar/Schoenhals 2008: 17). To protect Wu (and himself), Peng did not allow the article to be published in central media. However, Lin Biao authorised its publication in the Liberation Army Daily, which labelled the play "*an anti-party, anti-socialist and anti-Mao Zedong Thought poisonous weed which should be criticised*" (quoted in Barnouin/Yu 1993: 56). Peng's reaction was to authorise the 'February Outline Report', which he interpreted as being approved by Mao. However, when Peng eventually realised that Mao had been behind the whole affair, it was already too late, and his fate was sealed. Already in April 1966, Mao had declared the Outline to be a "*reversal of the relations between the enemy and ourselves, putting one into the position of the other*" (quoted in Barnouin/Yu 1993: 61). From that time on, there could be no academic or artistic issues separate from politics; both science and art were seen as inextricably tied to the class struggle.

31 In addition to constituting the remnants of the 'Yan'an roundtable', this group, later baptized 'Peng-Luo-Lu-Yang anti-party clique' was (coincidentally?) in charge of the state capital, propaganda work, as well as the day-to-day management of the PLA and the CCP's secretariat,. This comprehensive grasp of power was among the facts Lin listed as the evidence of the group's complicity in the 'bourgeois coup d'état'.

In such processes of singularisation, exclusive claims at generality are made through self-definition, which includes identity imputations and frames. The latter, in their turn, bring about counter-concepts discriminating against the 'excluded' (Koselleck 2004: 156). As a result, the excluded are imbued with negative identities. The People's Republic had seen a plethora of various rectification movements that purged the party of 'Rightists' and other 'reactionaries', and furthered the struggle against 'class enemies.'[32] Such labels functioned to deny the reciprocity of mutual recognition, which was not uncommon before and by itself would make the Cultural Revolution just another rectification campaign. However, this campaign went further than any previous one, and broke the rules that had guided the People's Republic since its foundation.

Overall, the Cultural Revolution appeared as a means to combat dangerous threats, beginning with counter-revolutionaries and spies and ending with insects, mice and sparrows, all of which required violent resistance in order to achieve victory and strengthening of the continuous revolution. Thus, political enemies were designated as 'evil demons and tigers'[33]—seemingly in accordance with the 'demonological paradigm'.

Indeed, the logic and legitimisation of the Cultural Revolution suggested that the struggle of the progressives had not ended with the victory in the civil war and the consolidation of the CCP; it was to continue with violent means for the sake of the revolution, and, eventually, for the sake of Mao Zedong as well. Thereafter, the securitisation of various top-party representatives allowed and facilitated their

---

32 These include the Anti Corruption Campaign 1951-1952 (三反, *sānfǎn*), the New Anti-Corruption Campaign 1953 (新三反运动, *xīn sānfǎnyùndòng*), the campaign to oppose corruption, speculation, waste, desintegration and bureaucracy 1953 (五反运动, *wǔfǎn yùndòng*), the Rectification Campaign 1954 (干部整风运动, *gànbù zhěngfēng yùndòng*), the anti-corruption campaign among tax-collectors 1955 (税务部门反贪污运动, *shuìwù bùméntānwūyùndòng*), another Rectification Campaign 1957 (整风运动, *zhěngfēng yùndòng*), and the Anti-Rightist movement 1957 (反右派运动, fǎnyòupài yùndòng). In fact, the Cultural Revolution was preceded by an anti-corruption campaign almost every year.

33 Tigers were understood as animals with a close connection to demons in traditional Chinese folk-beliefs (Haar 2002).

legitimate removal from the leadership, being a tool of the internal party politics addressing the fractional struggle in the CCP. In general, the security aspect of the Cultural Revolution served to construct an 'enemy' within society and justify legitimisation of a greater degree of violence against it than would have otherwise been normally permissible.

Beyond the use of violence, the securitisation arguments legitimised further 'transgression of rules'. In early 1966, the Communist Party had total control of Chinese politics and society. Yet, contrary to this objective and practice, the Cultural Revolution allowed unmediated attacks on party officials and institutions. With his policy of 'seizure of power', Mao seemingly ceded power to the 'streets' until the crackdown on the Red Guard movement in 1968.[34] These practices and

34 For example, the big character poster that accused the university authorities at Beida University of sabotage against the Cultural Revolution fit the bill of a political crime in 1966. Before this apparently spontaneous attack, the criticism of authorities had always been organised and initiated by the authorities themselves. Mao nonetheless hailed this as the "*first Marxist-Leninist big character poster*". Handling the Beida incident by so-called 'work teams', which was a standard way of dealing with such issues, contributed to Liu Shaoqi's undoing in 1967: despite the prior purge of Peng Zhen and the others, Liu Shaoqi failed to perceive the Cultural Revolution as a new kind of mass campaign, or even as a personal threat. By endorsing the formulations of the 'May 16th Circular' (Zhongfa 267), he did not recognise himself in 'the representatives of the bourgeois who had sneaked into the party'. Nor did Liu try to stick to his decision to send in the work teams to deal with the Beida incident, when Mao declared this action wrong. See MacFarquhar/Schoenhals (2008: 63-65) on the issue of work teams. The power seizure policy of 1967 revealed that Mao had a vision of an entirely new state machine (Barnouin/Yu 1993: 108-109); it seems that his plan was to replace the old state organs of power with 'revolutionary committees' based on 'great alliances of all revolutionary organisations' (MacFarquhar/Schoenhals 2008: 239). The youth of China had been brought up in a culture of violence (see Thurston 1990); as MacFarquhar/Schoenhals (2008: 131) note, the destruction the Red Guards was allowed to wreak went against the normally controlled and calibrated political violence of the party. Massive unrest seems to have been Mao's goal, as he toasted "*to the unfolding of nationwide all-round civil war*" on his 73th birthday in December 1966 (MacFarquhar/Schoenhals 2008: 155). In

the mass campaigns of terror directed to expose revisionists of one sort or another, put into question the whole basis of the political order. Indeed, it seems that Mao really desired to replace the whole system with personal leadership.

Mao was the self-evident hero of the revolution, but to get rid of the collective leadership was a radical measure. Since the Long March, the apex of CCP leadership had remained relatively stable; the changes that did take place were of secondary importance. The seventh Party Congress in 1945 had established the basic patterns of power structures within the party that lasted for some two decades until the Cultural Revolution was launched. In the mid-fifties, the revolution seemed consolidated enough to allow the party to focus on more practical matters.[35] In 1958 Mao had defined his own role in the party as retirement to the 'second front' of principal leadership. This would allow Liu Shaoqi and other leading cadres to take care of the day-to-day operations of the party and the state. However, in the early 1960s, Mao apparently chose to break with this fixed pattern (Barnouin/Yu 1993: 25); the targets of the purges implemented as a part of the Cul-

---

the end, most power seizures failed, and the military increased its influence, for example, in the form of the militarisation of ministries; even the 'revolutionary committees' retained the distinction between the revolutionary masses and the party vanguard (MacFarquhar/Schoenhals 2008: 160, 170-171, 240). The 'new political system' Mao had envisaged turned out to be a modification of the old, mainly manned by the officers of the PLA (MacFarquhar/Schoenhals 2008: 247).

35 In September 1956, the Eighth Party Congress had declared that the socialist system had been essentially established: "*contradictions among the proletariat and the bourgeoisie have been basically solved*" (Central Research Department of History 1986: 277), even though there still remained some remnants of the old system: "*the major contradiction still existing in China was between the advanced socialist system and the backward production forces in society*" (Barnouin/Yu 1993: 3). Mao began his 1957 speech on the correct handling of contradictions among the people by emphasising that "*never before has our country been as united as it is today.* [...] *The days of national disunity and chaos which the people detested are gone, never to return*". In this situation, the task for the party was the development of production and industrialisation, not the hunting down of class enemies or revisionists.

tural Revolution clearly went beyond the 'usual suspects' of rectification campaigns conducted before (MacFarquhar/Schoenhals 2008: 41).[36] By August 1966, with the Cultural Revolution at his disposal, Mao managed to replace the agreed-upon method of rule (collective leadership) with his personal power. In February 1967 the Central Committee ceased to function and was *de facto* replaced by the Cultural Revolution Group.

While Mao was in a position of unquestionable power even as the Cultural Revolution was launched, he nonetheless employed security arguments to purge the top leadership. It appears that even in Mao's case such purges, carried out without special justification, would have been viewed as an abuse of power. Securitisation of several top leaders of the party as counter-revolutionaries may be an example of practical application of the 'singularising' identities and concepts. From this viewpoint, it seems understandable that the Cultural Revolution was focused upon uncovering the "enemies" who had sneaked into the party.

The campaigns to ferret out counter-revolutionaries had major impact upon the political order of China and Chinese society at large. The template of counter-revolutionary threats was to be used over and over as one rectification campaign after another defined Chinese politics from the mid-1960s to the mid-1970s. The initial purge of Peng Zhen and the others paved the way to expelling Liu Shaoqi from the party. As the Cultural Revolution moved on, Liu was criticized in an increasingly systematic manner aimed at destroying his popular image. Mao's first heir apparent and a veteran of the revolutionary war, was vilified as a major revisionist and the leader of the 'dark headquarters' whose aim was the restoration of capitalism. On the whole, the allegiance to exceptionalist politics in China resulted in some 75 percent of the full or alternate members of the Eighth Central Committee suspected of being 'counter-revolutionaries' or some other form of 'traitor' (Barnouin/Yu 1993: 298; MacFarquhar/Schoenhals 2008: 273).

Once the threat discourse had served its purpose to legitimise the purge of the most highest-level party leaders, the effects of the Cultural Revolution spread outwards to finally encompass almost the entirety

36 In practice, it was only with Mao's authority that revisionists could be uncovered; only Mao could 'decide' who the revisionists at the top of the party were (MacFarquhar/Schoenhals 2008: 48).

of Chinese society; the mass campaigns involved over a hundred million people. For example, in January 1970, Mao linked the Soviet threat to class struggle within China, and launched yet another campaign—the so-called 'one smash, three antis'. The line of the campaign deemed it imperative to eliminate counter-revolutionaries before an alleged Soviet onslaught (Barnouin/Yu 1993: 190-191). Another example is the campaign which targeted the 'Counter-revolutionary May 16th Conspiracy' and affected 10 million people nationwide.[37] Current CCP historians present this campaign as an effort to 'ferret out' non-existent class-enemies (MacFarquhar/Schoenhals 2008: 221). If this is indeed so, then the fact that a non-existent conspiracy was successfully securitised, and this securitisation had such massive consequences, only highlights the point that an issue can become a security concern irrespective of the existence of a 'real' threat.

CONCLUSIONS

Reinhart Koselleck's notion of asymmetric concepts seems to operate in many societies and have various social functions. Here, I have explored one of the most drastic of such functions: the legitimisation of violence. I approached this theme through three takes on the counter-revolutionary, an institutionalised watchword for enemies in Mao's China.

The first take, Schmitt's notion of the political and the state of exception, was deemed too wide in respect to the more limited and specific functions of asymmetric political concepts in the tactics of politicking. Conversely, the second take—the so-called 'demonological paradigm' was shown to be too specific for the Chinese historical experience to suit broader comparisons of violence legitimation in various societies at different times. The third take of securitisation theory was deemed most developed and applicable for the kind of investigations attempted above. All in all, securitisation theory now can be

37 'Cleansing the class ranks' was a campaign that targeted some 36 million people. 750,000 to 1.5 million people were killed in order to 'resolve the contradictions between 'us' and 'the enemy'' (MacFarquhar/Schoenhals 2008: 253-262).

viewed as an appropriate device for studying the role of asymmetric political concepts in the legitimisation of violence.

REFERENCES

Agamben, G. [1995] 1998. *Homo Sacer. Sovereign Power and Bare Life*. Stanford: Stanford University Press.

Agamben, G. [1996] 2001. *Keinot vailla päämäärää. Reunamerkintöjä politiikasta*. Helsinki: Tutkijaliitto.

Agamben, G. [2003] 2005. *State of Exception*. Chicago & London: University of Chicago Press.

Balzacq, T. (ed.) 2010. *Securitization Theory. How Security Problems Emerge and Dissolve*. London and New York: Routledge.

Barnouin, B./Yu, C. 1993. *Ten Years of Turbulence. The Chinese Cultural Revolution*. London and New York: Kegan Paul International.

Buzan, B. 2008. The Changing Agenda of Military Security. In *Globalization and Environmental Challenges. Reconceptualizing Security in the 21st Century*, edited by H. G. Brauch/Ú. Oswald Spring/C. Mesjasz/J. Grin/P. Dunay/N. Chadha Behera/B. Chorou/P. Kameri-Mbote/P. H. Liotta. Berlin and Heidelberg and New York: Springer, 553-560.

Buzan, B./Wæver, O. 2003. *Regions and Powers*. Cambridge: Cambridge University Press.

Buzan, B./Wæver, O./Wilde, J. de. 1998. *Security. A New Framework for Analysis*. Boulder, CO: Lynne Rienner.

Central Research Department of History. 1986. 中共党史大事年报 Zhōng-Gòng dǎngshǐ dàshì niánbào [chronology of major events in the history of the CCP]. Beijing: Renmin Chubanshe.

Deng Xiaoping [1979] 1995. Uphold the Four Cardinal Principles. In *Selected Works of Deng Xiaoping, Volume II*. Beijing: Foreign Language Press, 168-191.

Dutton, M. 2005. *Policing Chinese Politics. A History*. Durham and London: Duke University Press.

Edelman, M. [1964] 1972. *The Symbolic Uses of Politics*. Urbana, Chicago and London: University of Illinois Press.

Haar, B. J. t. 1997. China's Inner Demons. The Political Impact of the Demonological Paradigm. *China Information* 11 (2): 54-88.

Haar, B. J. t. 2002. China's Inner Demons. The Political Impact of the Demonological Paradigm. In *China's Great Proletarian Cultural Revolution. Master Narratives and Post-Mao Counternarratives*, edited by W. L. Chong. Lanham & Boulder & New York & Oxford: Rowman & Littlefield, 27-68.

Heinzig, D. [1998] 2004. *The Soviet Union and Communist China 1945-1950. The Arduous Road to the Alliance*. Armonk: M.E. Sharpe.

Hughes, C. R. 2006. *Chinese Nationalism in the Global Era*. London: Routledge.

Huysmans, J. 2006. The Politics of Insecurity. Fear, Migration and Asylum in the EU. London and New York: Routledge.

Koselleck, R. [1979] 2004. The Historical-Political Semantics of Asymmetric Counterconcencepts. In *Futures Past, on the Semantics of Historical Time*, edited by R. Koselleck. New York: Columbia University Press, 155-191.

Lenin, V.I. [1917] 2004. *The State and Revolution*. Whitefish: Kessinger Publishing.

林彪 Lin Biao. 1966. 在中央政治局扩大会议上的讲话. In 美国《中国文化大革命文库光盘》编委会 编纂 香港中文大学• 中国研究服务中心 制作及出版 (The Cultural Revolution Database), edited by 宋永毅.

MacFarquhar, R./Schoenhals, M. [2006] 2008. *Mao's Last Revolution*. Cambridge and London: The Belkap Press of Harvard University Press.

Neocleous, M. 2007. Return the Gift. Violence, Emancipation, Critique. *Keynote at Critical Approaches to Security in Europe III. Emancipation, resistance, Violence Joint Seminar for Younger Scholars Programme*. University of Southern Denmark 29th November - 1st December 2007.

Palmujoki, E. 1995. *Revolutionary Pragmatism and Formal Marxism-Leninism. An Analysis of Vietnam's Foreign-Policy Argumentation from the Fall of Saigon to the Collapse of the Socialist World System*. Tampere: University of Tampere.

Paltemaa, L./Vuori, J. A. 2006. How Cheap is Identity Talk? A Framework of Identity Frames and Security Discourse for the Analysis of Repression and Legitimization of Social Movements in Mainland China. *Issues and Studies* 42 (3): 47-86.

Perry, E. J./Selden, M. (eds) 2003. *Chinese Society. Change, Conflict and Resistance. 2nd Edition.* London & New York: RoutledgeCurzon.

Pye, L. 1992. *The Spirit of Chinese Politics. 2nd Edition.* Cambridge: Harvard University Press.

Radtke, K. W. 2008. Security in Chinese, Korean, and Japanese Philosophy and Ethics. In *Globalization and Environmental Challenges. Reconceptualizing Security in the 21st Century*, edited by H. G. Brauch/Ú. Oswald Spring/C. Mesjasz/J. Grin/P. Dunay/N. Chadha Behera/B. Chorou/P. Kameri-Mbote/P. H. Liotta. Berlin and Heidelberg and New York: Springer, 203-220.

Rinehart, J. F. 2006. *Apocalyptic Faith and Political Violence. Prophets of Terror.* New York: Palgrave MacMillan.

Schmitt, C. [1933] 1996. *The Concept of the Political.* Chicago and London: The University of Chicago Press.

Schmitt, C. [1922, rev. ed. 1934] 2005. *Political Theology, Four Chapters on the Concept of Sovereignty.* Chicago and London: University of Chicago Press.

Shek, R. 1990. Sectarian Eschatology and Violence. In *Violence in China. Essays in Culture and Counterculture*, edited by. J. N. Lippman/S. Harrell. Albany: State University of New York Press, 87-114.

Snow, D. A./Benford, R. D. 1992. Master Frames and Cycles of Protest. In *Frontiers in Social Movement Theory*, edited by. A. D. Morris/C. McClurg Mueller. New Haven and London: Yale University Press, 133-155.

The Polemic on the General Line of the International Communist Movement. 1965. Peking: Foreign Language Press.

Thurston, A. F. 1990. Urban Violence During the Cultural Revolution. Who Is to Blame? In *Violence in China. Essays in Culture and Counterculture*, edited by. J. N. Lippman/S. Harrell. Albany: State University of New York Press, 149-174.

Vuori, J. A. 2008. Illocutionary Logic and Strands of Securitisation. Applying the Theory of Securitisation to the Study of Non-Democratic Political Orders. *European Journal of International Relations* 14 (1): 65-99.

Vuori, J. A. 2010. Religion Bites. Falungong, Securitization/Desecuritization in the People's Republic of China. In *Securitization Theo-*

ry: How Security Problems Emerge and Dissolve edited by T. Balzacq. London and New York: Routledge, 186-211.

Wæver, O. 1995. Securitization and Desecuritization. In *On Security*, edited by R. D. Lipschutz. New York: Columbia University Press, 46-86.

Wæver, O. 2008a. Peace and Security. Two Evolving Concepts and Their Changing Relationship. In *Globalization and Environmental Challenges: Reconceptualizing Security in the 21st Century*, edited by H. G. Brauch/Ú. Oswald Spring/C. Mesjasz/J. Grin/P. Dunay/N. Chadha Behera/B. Chorou/P. Kameri-Mbote/P. H. Liotta. Berlin and Heidelberg and New York: Springer, 99-112.

Wæver, O. 2008b. The Changing Agenda of Societal Security. In *Globalization and Environmental Challenges: Reconceptualizing Security in the 21st Century*, edited by H. G. Brauch/Ú. Oswald Spring/C. Mesjasz/J. Grin/P. Dunay/N. Chadha Behera/B. Chorou/P. Kameri-Mbote/P. H. Liotta. Berlin and Heidelberg and New York: Springer, 581-594.

Wang, Z. 2008. National Humiliation, History Education, and the Politics of Historical Memory. Patriotic Education Campaign in China. *International Studies Quarterly* 52 (4): 783-806.

Wittgenstein, L. [1953] 2001. *Philosophical Investigations*. 3rd Edition. Oxford: Blackwell Publishing.

Yang, C. 1961. *Religion in Chinese Society. A Study of Contemporary Social Functions of Religion and Some of Their Historical Factors*. Berkeley: University of California Press.

中发 [1966] 267号 [Zhongfa 267, May 5th, 1966]. 中国共产党中央委员会通知及原件附件之二 [CCP Circular with the original appendix 2]. In 美国《中国文化大革命文库光盘》编委会 编纂 香港中文大学• 中国研究服务中心 制作及出版 (The Cultural Revolution Database), edited by 宋永毅.

Žižek, S. 2002. *Did Somebody Say Totalitarianism? Five Interventions in the (Mis)use of a Notion*. London: Verso.

'*We* are the Barbarians'

Thomas Hobbes, the American Savage and the Debate about British Antiquity

PHILIP MANOW

I

It might be important to note that Koselleck's 1975 article on the "historisch-politische Semantik asymmetrischer Gegenbegriffe" (Koselleck 1989) is an elaboration of a lengthy endnote that appeared earlier in his dissertation. In this instant classics Koselleck discusses the nexus between the national pacification of the (religious) civil war through the emerging absolutist state in the 17th century and the era of state wars that begins at the same time (Koselleck 1973: 171-172). Koselleck points to this nexus between the internal peace and the international 'state of nature' (i.e. the 'continuall warre' between states as *magni homines*) in the context of his discussion of Hobbes. He ascribes to the latter the argument that once the Leviathan has managed to overcome the civil war of all against all, the internal aggression is directed outwards, turns international.

This argument comes in two different, slightly contradictory variants, and Koselleck does not seem to favor one over the other. The first variant sees the end of the civil war *as a precondition*: since the emerging national states have 'no common power to fear' (and after the Reformation accept no common moral or religious standard, or authority anymore) a nation, to be capable of competing in an emerging international order ruled by nothing than brute force, *needs to over-*

come internal strife. To avoid conquest, a nation must establish effective sovereignty, which means surrendering individual sovereignties to one man or 'an assembly of men'. In Leviathan, Hobbes stresses this point over and over again in various formulations: One has to build a commonwealth on consensual or institutional grounds to avoid the commonwealth based on conquest or acquisition (Tuck 1998: xl).[1]

The second variant establishes a different nexus, and has a rather Schmittian flavor to it. National unity can be achieved mainly *by way* of identifying and fighting a common enemy (Schmitt: 'Der Feind ist unsere eigene Frage als Gestalt'; quoted from Laak 2002: 142, footnote 40). In this sense, war with other nations *is a means* to overcome internal strife. Koselleck finds both variants in Hobbes. He quotes Leviathan: "And Law was brought into the world for nothing else, but to limit the naturall liberty of particular men, in such manner, *as they might not hurt, but assist one another*, and joyn together, against a common Enemy" (Leviathan II: 26; quoted in Koselleck 1973: 171, endnote 100). And adds to this quotation: "with the end of the civil war, with the inner consolidation of states, war turns outwards and many absolutist thinkers see in war a permanent institution *to avoid civil war*" (Koselleck 1973: 34, my emphasis).

Koselleck's main argument in these paragraphs goes as follows: "The end of civil war and the taming of war in the form of the pure war between nation states (reiner Staatenkrieg) are two corresponding phenomena" (Koselleck 1973: 34). And it is exactly this *correspondence*—a notion which leaves it open whether causality runs from in-

1 "[T]hose who subject themselves to another through fear either submit to the person they fear or submit to some other whom they trust for protection. [...] The first mode has its beginning in natural power, and may be said to be the natural origin of the commonwealth; the latter originates in the determination and decision [a consilio & constitutione] of the uniting parties, and that is the origin by design [origo ex instituto]. Hence there are two kinds of commonwealths: one kind is natural, like the Paternal and the Despotic; the other is the kind of commonwealth which is by design [institutivum], and which may also be called political. In the first case a Lord [Dominus] acquires citizens for himself by is own will; in the second, the citizens impose a Lord upon themselves by their own decision, whether that be one man or one group of men with sovereign power [summum imperium]" (*On the Citizen* V: 12).

ternal peace to external war or vice versa—that is stabilized by asymmetrical counter-concepts ('asymmetrische Gegenbegriffe'), we are told. What was in ancient times the asymmetrical contrast of Greeks versus barbarians, was in medieval times the contradistinction between Christians and pagans: "A spatial contradiction always gives direction and difference [*Gefälle*] to a political contradiction" (Koselleck 1973: 34).

Hobbes's times witness only one small but important innovation: the acknowledged legal status of the enemy in the emerging international law (in particular in the martial law). The nascent *ius gentium europaeum*, which emerges in the late 16th, early 17th century with Hugo Grotius and John Selden, comes into being by working with a new, geographic *Gegenbegriff*: Christian, civilized Europe (*imperium*) contrasts itself to pagan, savage America (*libertas*). Located beyond the so-called "amity-line", America demarcates a space of the state of emergence—a space where everything is allowed, where there is no law and no sin (cf. Schmitt 1950; Damler 2008). And it this extra-legal status of both the American savage and this zone beyond the amity-line that helps to bring about the mutual legal recognition between the competing European nations. The various 'civilized nations' discover their commonality in their "common" struggle for the oversea territories through the 'otherness' of American, Asian and African savages they conquest. This is quite a new and modern definition of 'us vs. them': the 'we' is defined by the equality of the mutual legal recognition as accepted, rightful enemies (*justus hostis*) (Schmitt 1997)—whereas the war against the other enemy—the savage—is waged without rules.[2]

2 Of course, Europeans *were* aware of a difference between a war between Christian kings and a war of Christians with the heathens before: the one was 'contained' by law, and the other was without any rules. The novelty was the fact that the 'contained' war between nations was now simply based on their status as sovereign nations, whereas the old ordo of the European Respublica Christiana—the foundation of the earlier distinction—had dissolved. This development is masterfully reconstructed in (Schmitt 1997)—to which both Koselleck (1973) and Koselleck (1989) obviously owe a great deal.

In this contribution I would like to go back to Koselleck's discussion of Hobbes. I would like to highlight the critical influence that the American discoveries had on Hobbes's political ideas. In particular I will argue that in the European colonial experience two of Koselleck's asymmetrical concepts—*Christian* vs. *heathen* and *Civilization* vs. *Barbarism*—overlap: while both introduced distinct temporal orders, only one of these orders had been pointed out by Koselleck himself. The temporal order inherent in the Christian/heathen contradistinction is this: the heathens of today shall be the Christians of tomorrow (Koselleck 1989: 239). In this projection into the future the Christian 'we' has a positive connotation. The second contradistinction implies a new and different conception of time: in times past we were as savage as the inhabitants of America are today—we were barbarians, too. This implied a projection into the past in which the 'we' suddenly acquired a questionable moral status. But, taken together, both projections introduced a new understanding of historical progress, although this progress was perceived as precarious, because of the permanent danger to return into the state of nature. Hobbes warned that society may always 'degenerate into civil war', and that man can become a savage again (*Leviathan* XIII: 11).

Important, however, is that the layering of these two asymmetrical concepts with their accompanying temporal orders destabilized the established asymmetrical distinctions of 'us versus them'. In particular, if seen from a Protestant perspective, the one concept undermined the other: from the Protestant standpoint, Catholic Spaniards in the colonies behaved more barbarically than American savages. Bartolomé de Las Casas' *Brevisima Relación de la Destruccíon de Las Indias* (1551), a report of "the shocking cruelty shown to the Indians by the Spaniards", became known in England in the 1590s (Kendrick 1950: 123). Given that with the reformation and the ensuing religious wars the very Christian notion of 'we' became bitterly contested, its implicit assumption ofl supremacy began to erode. In Luhmann's terms, in the context of European religious schism and the American discoveries we observe the re-entry of the distinction between Barbarism and Civilization into the distinction between Christians and Pagans.[3] This re-entry,

3 One may wonder why the Reformation did not result in a new pair of asymmetrical concepts in which 'we the Protestants' (or 'we the Catholics') stood for the true religion and all others represented a false one. Ap-

I would claim, undermined the dominance of the religiously based antithetical notions.

I want to argue that all this has had profound implications for the emerging collective self-understanding. It is reflected in the modern political philosophy that began to take shape around this time and found its masterful expression in Thomas Hobbes' work. In the light of their own description of the others (Indians) as barbarians, Brits of the late 16th - early 17th century began to re-consider ancient sources that had described themselves as barbarians. In the light of their own description of the (American) other, Europeans began accepting the ancient descriptions of themselves as others. In learned humanist circles, the writings of Caesar, Pliny, Herodian, Claudian about the British people were now accepted as sources for a new national self-image, a less heroic and mythical and a more 'ethnological' one than that which had previously dominated British collective self-understanding (Kendrick 1950; Kidd 1999; Levy 1967). This paved the way for the modern view that *all* societies originate in a *universal* state of nature. The modern argument of 'ethnological time'—the possibility of seeing in the savage one's own ancestor—was born. This introduced the very new idea of historical progress, which—in the light of what has been said before—probably needs to be predated: to me it does not seem to

parently, given the ever-changing coalitions between different nations in the emerging world of European statehood of the 16th and 17th century the confessional divide could only temporarily stabilize national identities and the nation became the dominant locus of identity formation. It was only in late Elizabethian times that the English nation developed a self-understanding of a 'Protestant nation' in its conflict with Habsburg Spain and Austria (Collinson 1988; Scherneck 1996). But already the alliance with Catholic France, and later the war with the Protestant Netherlands collided with any purely confessional self-understanding and self-affirmation. Moreover, in the wake of the religious conflicts it became also clear that denominations clearly transgressed national boundaries (Koselleck 1975). With respect to the Huguenot French and Calvinist Dutch exiles who found refuge in England, William Bradshaw wrote: "Touching the word foreign, those Churches being all of the same household of faith that we are, they are not aptly called foreign" (quoted from Collinson 1988: 16).

be a bourgeois invention of the 18th century but rather a European one of the 17th century.

My contribution proceeds as follows. First I will briefly point to the frontispiece of Hobbes' *De Cive* published in 1642. I will show how much this title page confronts us with a detailed, subtle argument about Europe and America, *imperium* and *libertas*, present and past, civilization and barbarism, and how it displays a novel argument about a process of civilization and its dangers. I will also demonstrate how much Hobbes in this title page drew on contemporary travel reports and their images of the American savage. I will continue by emphasizing that Hobbes's argument, which basically equated "the experience of savage nations that live at this day" with the "histories of our ancestors" (*The Elements of Law* XIV: 12), was a contribution to a broader contemporary debate about British antiquity (Kendrick 1950; Kidd 1999; Levy 1967). I will briefly sketch this debate and its implications for the emerging new national self-understanding, which finally put away with the mythological narratives of the foundation of the British nation and instead accepted the new view about the barbarian origins of one's own society. In the light of this discussion I will then finally return to Koselleck's notion of asymmetrical concepts.

II

The title page of Hobbes' *De Cive* Hobbes furnishes us with a nuanced argument about America as an example of the state of nature that had been shared by all in earlier times (see Figure 1). The title page is divided into three parts, labeled *religio*, *imperium* and *libertas*. The three parts of the picture correspond to the three books of *De Cive*, which are 'government' (Book 1), 'liberty' (Book 2), and 'religion' (Book 3). In the upper third of the picture, hovering over the entire exposition, we see a scene of the Last Judgment, where devils capture the sinners and lead them to hell on the right, while Jesus and angels help the pious souls in their ascent to heaven on the left. Below this scene we see the contrast between the left and the right, expressed in conceptual pairs: Europe vs. America, *imperium* vs. *libertas*, culture vs. nature, civilization vs. savagery, peace vs. violence, and, finally, 'us' vs. 'them'. In conformity with "the iconographical tradition of contrasting America and Europe as ideal female types" (Hoekstra 2007: 113), both *imperium* and *libertas* are pictured as two women standing on pedestals: in the first case, it is a beautiful young woman with crown, sword and scale; in the second case—a barely dressed, old, ugly and grimly looking women with a longbow on her left and an arrow pointing downwards on her right (note that Imperia's sword, on the contrary, points upwards to the heaven (cf. Skinner 2008: 99; Bredekamp 1999: 144). Further, we see the peaceful cultivation of land in the background on the left contrasted with savagery, man hunt and cannibalism on the right. The "native American standing before scenes of savage violence symbolizes the Hobbesian state of nature" (Collins 2005: 62-63; cf. Hodgen 1964).

Figure 1: The "De Cive" title page (1642)

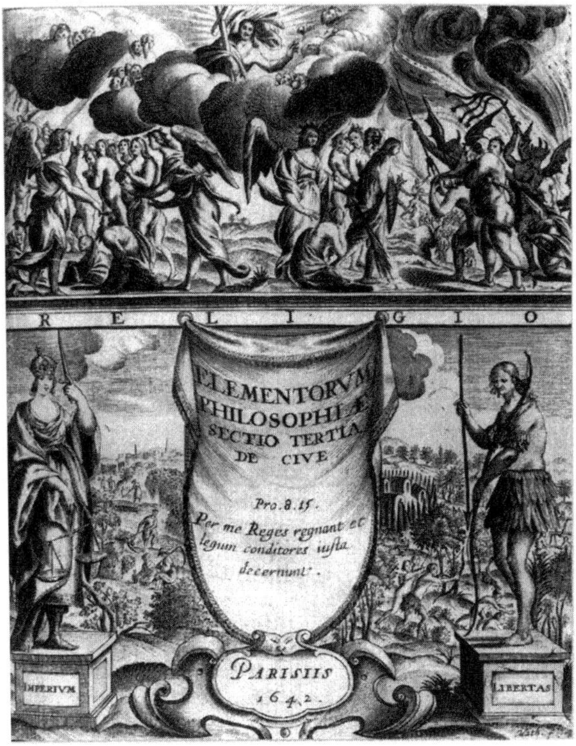

Source: Skinner 2008: 100

Clearly, this is a classical asymmetrical visualization, but as we will see, it is complicated by a subtle argument that tends to undermine any self-assuring contrasts between 'us', the civilized Europeans, and 'them', the wild Americans. On the one hand Hobbes argues that in former times, before we all subdued to a sovereign, we too lived in such a cruel, brutal state of nature. On the other he emphasizes that we may always degenerate to this extremely unpleasant, poor, nasty and brutish state.[4] Consequently, the title page of *De Cive* draws a timeline between *imperium* and *libertas* that can be read *both* from right to left *and* from left to right: the *imperium*, or commonwealth, evolved out of the state of nature, represented by the figure of the American Indian to the right, but it always might degenerate back into this state of nature in case of civil war. The picture's temporal ambiguity in itself is a visual expression of the new understanding of history, progress, and human agency that begins to gain ground in the early 17th century (cf. Boehm 1969).

Of particular importance for our context is the fact that the *De Cive*-frontispiece contains a couple of direct quotations from De Bry's edition of Thomas Harriot's American travel report—the 'new found land of Virginia', written in 1585. In the first volume of the famous America series issued by the de Bry publishing house, which contained Harriot's report, Theodore de Bry had explicitly contributed to the contemporary debate about the barbaric origins of the British nation by adding five pictures of ancient Britons to Harriot's report 'in order to show that in times past the inhabitants of Britain were as savage as those of Virginia' (see Smiles 2009). Now Hobbes quotes the images of de Bry in the frontispiece of his *De Cive*. Its title page displays numerous direct quotations from the first volume of de Bry's 'grandes voyages' (see Figure 2). These borrowings are common knowledge in the literature on Hobbes (Skinner 2008: 101; Tuck 1996: XXV, footnote 29; Corbett/Lightbown 1979: 224-225; Malcolm 2002: 61, footnote 34), but as of now have remained unnoticed in the rich literature on the painter John White (Feest 2007; Sloan 2007, 2009; Hulton 1984; Hulton/Quinn 1964), or on the engraver and printer The-

4 "[I]t may be perceived what manner of life there would be, where there were no common power to fear; by the manner of life, which men that have formerly lived under a peaceful government, use to degenerate into, in a civil war." (*Leviathan* XIII: 11)

odor de Bry (Greve 2004; Duchet et al. 1987; Burghartz 2004) and their visual legacy.

Figure 2: The title page of the 'briefe and true report of the new found land of Virginia' (1590)

Source: Harriot 1972: 1

The most obvious case of a direct visual quotation is the Indian figure on the title page of Volume I, which itself refers to a plate inside the report showing 'A weroan or great Lorde of Virginia'. Besides, the female imperium figure of Hobbes's *De Cive* also seems to cite the frontispiece de Bry's first volume: in particular, the unusual way in which she holds her right arm with the scale reminds of de Bry's Indian female figure to the right (see Figure 2). Furthermore, in de Bry's frontispiece (as in the *De Cive*-title page) the Indian points the arrow downwards, which, in its turn, is a liberal interpretation of John White's watercolor on with this engraving was based (Figure 3, see below). Other visual features of the title page of *De Cive*—for instance, the palisade village in the background—refer to the illustrations to Harriot's 'brief report' as published by de Bry, too. All in all, "Hobbes's general composition appears to owe much to de Bry" (Skinner 2008: 102; see for additional evidence Manow 2011).

De Bry's prestigious edition project, finally comprising 14 richly illustrated volumes, was a publication "with a maximum European impact" (Hulton 1984: 17). It became "widely known and copied throughout Europe", and was an "extremely successful" publication (Kuhlemann 2007: 79, 82; Sloan 2007, 2009) "assured of a long life in European libraries" (Quinn 1979: 139; Greve 2004). The first volume of this project, Harriot's report, had been published in Latin, German, French and English and later appeared in two German (1600, 1620) and two further Latin editions (1608, 1634), whereas the other 13 volumes of the series were published in Latin and German, only. It is generally consented that the edition project had a formative, authoritative influence on 17th century European perception of the oversea territories and their inhabitants (Kuhlemann 2007: 79; Gaudio 2008). Here, however, I would like to point out their contribution to the self-perception of the conquering nations like Britain, i.e. on their formative impact on the understanding of 'we' rather than 'them' (cf. Smiles 2009).

In this context it is of relevance that in the first America volume Theodor de Bry integrated five engravings allegedly showing British inhabitants of ancient times—the so-called Picts. De Bry explains that by including of these pictures of the 'ancient Britons' into the volume he intended "to showe that in tymes past the Pictes of great Bretainnie were as sauvage as those of Virginia" (quoted from Hulton 1984: 130). With the help of a classical trope de Bry tries to convince his readers that the pictures are authentic: the plates are embellished with titles such as 'The *true* picture of one Pict', 'The *true* picture of a woman Pict' etc. He is at pains to ward off any suspicion that these pictures might have been the product of sheer fantasy, claiming that they have been 'fownd […] in a oolld English chronicle'. By this, he imitated those very authors against whom these pictures were primarily directed—namely the writers who had constructed fanciful accounts of a British pre-history purportedly reaching back as far as 2,000 or even 3,000 years. Such lavish genealogies were typically constructed by referring to some long lost but suddenly rediscovered old manuscripts, some as of yet unknown *vetustissimus liber* that miraculously proved the very long genealogies of the British nation reaching back at least to Troja (as in Geoffrey of Monmouth's *Historia Regum Britanniae*, around 1148), if not straight to Noah's ark (as in John Bale's *Illustrium majoris Britanniae Scriptorum* [...]., 1548). "By relating American Indians and their lifestyle [...] to earlier European civilizations" (Kuhlemann 2007: 85), the images of the Picts were primarily directed against these myths of noble and ancient descent.[5] In the heated contemporary debate about British antiquity, de Bry thereby took a very pointed stand together with Camden, Raleigh, Oretelius and those others who disputed the veracity of the elaborated genealogies of British kings and the British nations and denounced them as mythic fabrications concocted in order to claim dynastic supremacy in the fierce status rivalry among European royal houses.

De Bry's visual argument about the savage origins of the Britons finds its direct reflection on the title page of Hobbes' *De Cive*, as well as in the argument about the barbarian origins of the 'now civilized nations' that Hobbes had made several times in his *Elements of Law, De Cive*, and *Leviathan*. Whenever Hobbes intended to give examples of the state of nature, he invoked both the American present and the European past and thereby forged a link to the asymmetrical conceptual pair 'Barbarians' *vs*. 'Hellenes'. In the *Elements of Law* we are told that the state of nature is exemplified by "the experience of savage nations that live at this day, and by the histories of our ancestors, the old inhabitants of Germany and other now civil countries, where we find the people few and short-lived, and without the ornaments and comforts of life" (*The Elements of Law* XIV: 12). Similarly, in *De Cive* Hobbes argues:

---

5 In fact, Theodore de Bry gives us a very interesting, if heterogeneous, visual account of history, since the very first plate in his edition of Harriot's report shows Adam and Eve in paradise. This way the engravings documenting the life of the Carolina Algonkians are, so to speak, embedded in a broader, if incoherent, story. De Bry's unoriginally begins his narrative with 'biblical time', but in the end comes up with a secular argument about the 'ethnological time' relating the life of the Indians to those of the 'inhabitants of Britain of ancient times'. So Theodor de Bry remains stuck in the middle between the traditional and the modern, oscillating between the religious and the secular historiographical accounts.

"One may easily see how incompatible perpetual *War* is with the preservation of the human race or of individual man. [...] The present century presents an example of this in the Americans. Past centuries show us nations, now civilized and flourishing, whose inhabitants then were *few, savage, short lived, poor and mean*, and lacked all the comforts and amenities of life which peace and society afford" (*On the Citizen* I: 13, emphasis added).

Familiar as this argument about our savage origins may sound today, it was far from being universally accepted when Hobbes spelled it out. Hobbes, as de Bry before him, assumes a very clear and radical position in the polemic between the new 'scientific, 'rational' or ethnological understanding of the national past and those narratives relating British society to either biblical or antique origins. Here is not the place to describe in much detail the complex history of the debate on British antiquity reaching back to Geoffrey of Monmouth's 'Historia Regum Britanniae', in particular because it has already been masterfully summarized by other scholars (Kendrick 1950; Shields 2007; Levy 1967).[6] What is of importance here is rather the fact that this dispute was barely settled in the mid-17th century when Hobbes wrote his major works, regardless of the major blows dealt to the "old" heroic historiography by John Speed's *History of Great Britaine* (1611) or William Camden's *Britannia* (1586, after four Latin editions translated into English in 1610). Although these books "fatally wounded" the

6 Smile gives a very brief summary of the 'glorious history' of England contained in Monmouth's *Historia Regum Britanniae* (around 1148). It begins "with the fall of Troy and the journey of Aeneas' great-grandson, Brutus, to the country then called Albion. When the colonists arrive the giants inhabitating the land are defeated, among them Gogmagog, and the Trojans rename the country Britain after Brutus. A long-lasting dynasty is then established among whose rulers may be found Lear and Cordelia, Cymbeline and Arthur. And on the banks of the Thames a capital city, Trinovantum, Troynovant or New Troy, is founded, which will later be renamed London after King Lud. The chronological span of ancient British history, in Geoffrey's telling, runs from the sack of Troy, c. 1240 BC, to the death of Cadwaller in AD 689. The overall import of the Galfridian account is that Britain's civilization antedated that of Rome, that Britain acquired through conquest an extensive continental empire and that British history was as glorious as anything found in Classical annals" (Smile 2009: 109).

fantastic narratives about British Antiquity, the broader reception of the new scholarship was still limited (Smiles 2009: 110-111). In particular, the fables of the ancient British origins remained popular elements in theatre plays (e.g. Shakespeare's *King Lear*), epic poems (e.g. Drayton's *Poly Olbion*, 1612), pageants, and, more generally, in the public display of royal splendor and sovereignty.

In fact, the myths and fables about the noble origins of the British nation were still alive in the dynastic-centered tales that were employed in early Stuart-England for the purpose of political legitimation. Myths of the nation's noble descent were used to inspire patriotic sentiment and back the claims of supremacy of the British monarchy over the other European royal houses. As a prime example for late Tudor-times could be mentioned Spenser, who wrote, praising Elizabeth: 'Begin, O Clio, and recount from antiquitie/ My glorious Soueraines goodly ancestrie", or: 'the famous auncestries/ Of my most dreaded Soueraigne I recount/ By which all earthly Princes she doth farre surmount' (Spenser 1596, quoted from Uhlig 1996: 262, 272). Given their public functions, these myths did not vanish overnight in the light of the new scholarly criticism represented by Camden, Raleigh, Hakluyt, Speed and others. As Kendrick summarizes: "the struggle to apply to the British history the critical machinery of a new scholarship did not speedily result in an easy and inevitable triumph of the Renaissance mind, but was one that had to be sustained for over a century against a formidable deadweight of contrary opinion jealously preserved and defended by reactionary scholars of great learning and repute" (Kendrick 1950: 77).

This shows that Hobbes intervened in an ongoing debate when he stressed the barbarian origins of the 'now civilized nations'.[7] For Hobbes—as for many others of the 'new critical school'—the prime proof for the barbarian origins of the British nation were the American savages. The American travel reports therefore were critical in the establishment of a new world view, since due to the vested interests in upholding the fantastic histories of British antiquity it "needed a shock

---

7 And if we go beyond the British debate, works like Grotius's *De Origine gentum Americanarum* (1643) or La Peyrère's *Praeadamitiae* (1655), which tried to reconcile the American discoveries with biblical authority, also demonstrate how much new and old genealogies still struggled with each other for dominance (Grafton 1992: 210-211).

[…] to make us admit that we had once been noble savages". And "it seems probable" that this "shock was provided by the Red Indians" (Kendrick 1950: 121). To this New World shock and its reverberations in Hobbes's work I would like to turn now.

III

To see how the old Hellenes/Barbarians-distinction in its New World-variant helped to re-discover an insight that had got lost in medieval times, namely that "Hellenes in old times followed the same customs that Barbarians follow today" (Thukydides, quoted from Koselleck 1989: 222-223),[8] the case of John White's pictures depicting ancient Brits is instructive. Sally Birch in a fascinating essay has recently shown that John White's paintings of the Picts were in all likelihood copies of the painting 'A Young Daughter of the Picts' attributed to Jacques Le Moyne de Morgues (Birch 2009). What is more, John White apparently relied upon this painting not only in the five water-colors of the Picts on which Theodor de Bry's engravings in the first America-volume were based, but also in the watercolor 'A weroan or great Lorde of Virginia', which then was the source for the title page of de Bry first America-volume and of Hobbes's *De Cive* (Birch 2009: 92-93). Closer inspection shows that the picture of the Indian chief was a "later and sloppier derivation from the Warrior template" (Birch 2009; see Figure 8), while this template itself was a copy of Le Moyne's 'Young Daughter'. When de Bry traveled to London in the late-1580s, he received from John White—probably with the help of Richard Hayklut—the watercolors that the painter had produced during Harriot's Virginia-expedition in 1585, and de Bry, in all likeliness, commissioned from White further pictures of early Brits to complete the first volume of his 'America'-series. Many details make it quite likely that the painting 'Young Daughter' served as a model for these additional pictures. De Bry was well acquainted to Jacques Le Moyen, both were "united by their devout Huguenot beliefs" (Birch 2009: 94). Le Moyne, however, was at this point already too ill to contribute new material to de Bry's publication project, and he was known to be ex-

8 In this context it seems worthwhile to note that Hobbes's first publication in 1627 was a translation of Thukydides's Peloponnenian War.

pensive (Birch 2009: 94). The pictures' genealogy therefore obviously starts with Le Moyne's painting—a purely fantastic vision of the British Picts, then leads to a first adaption made by John White ('Warrior Pict with severed Head'), which White himself used later as a template for his other drawings of the ancient Picts and "at least for one [depiction] of the Indian subjects" (Birch 2009: 94), namely for the 'A weroan or great Lorde of Virginia' picture. Finally, this last derivation of an Indian subject served de Bry as a model for his engravings on the title page and inside of his first volume of the 'America'-series (see Figure 2, above and Figure 5, below)—the one that was quoted by Hobbes (see Figure 1).

Figure 3: Pictish Woman, John White

Source: Hulton 1984: plate 65

In other words, John White with his Picts did not simply try to invent a "European parallel to the painted and tatooed decoration he found among the Algonquian Indians" (Hulton 1984: 185). To some extent, his imagination worked in the opposite direction: he visualized the Indians with reference to a purely fictive vision of the Britons of early times. This helps us to explain some of the classic features of John Whites's pictures, which were claimed by their author to have been 'counterfeited according to the truth', but nevertheless clearly point at the highly specific European iconographical tradition. Against the background of the pictures' protracted genealogy it is far less surprising to discover some classic elements such as the 'Renaissance elbow' in allegedly ethnologically authentic accounts of the Algonquian Indians (Pratt 2009: 35; see Figure 5).

Figure 4: Man of the Pictes, John White

Source: Hulton 1984: plate 67

Figure 5: A weroan or great Lorde of Virginia

Source: Theodor de Bry 1590

Figure 6: 'An Indian weroan or great Lorde of Virginia', superimposed with a tracing on 'Warrior Pict with severed head'

Source: Birch 2009: 93

IV

Early anthropology (Hodgen 1964; Lantucci 1972) was crucial for the new understanding of society's origins that began to emerge in the late 16th, early 17th century. It has been shown above how important it is to place Thomas Hobbes's political philosophy within the broader contemporary debate about the pre-history of the British nation, and how much his concept of the 'state of nature' owes to the radically new understanding of mankind's beginnings that began to take shape around this time. Being the radical thinker that he was, Hobbes instantly gets rid of all mythological, heroic national narratives and postulates a universal original state—that of brutal barbarity. His argument was heavily influenced by the colonial experience, leading to the new central insight that Locke famously formulated only a little later: 'In the beginning, all the world was America' (Locke 1988, *Second Treatise of Government:* § 49). The paintings and engravings of John White and Theodor de Bry, of which Hobbes was acutely aware and which he cited in the frontispiece of *De Cive*, provide one of the most stunning visual arguments about the emerging new understanding of the nexus between the American present and the European past. I would like to conclude with two rather speculative remarks—one about the 'epic confrontations' between East and West that lurk behind the various asymmetrical concepts shaping our political semantics, and another about the new understanding of historic and societal time that begins to take shape in Hobbes's times.

(1) All three of Koselleck's asymmetrical pairs—Hellenes vs. Barbarians, Christians vs. Heathens, Mensch vs. Unmensch—are based on a geographic coding: West versus East. Homer's *Iliad* already deals with a conflict between Greeks against Trojans, or Europe against Asia—an opposition that is repeated in Herodotus' *Histories* and in Vergil's *Aeneid*. In Medieval Europe the contrast between Christians and the *infideli* is also presented as a conflict between West and East: the heathens live in Turkey, Persia, middle and far East. Even the Nazi dichotomy Mensch vs. Unmensch has been based on the understanding that the *Un-* or *Untermensch* lives in the East.[9] This East vs. West di-

9 In fact, the practice of mass persecution and mass extinction began during the Russian campaign and, as we know today, was probably developed as a military strategy aimed at exterminating the major part of the 'useless, in-

chotomy looks less surprising when one realizes that it is a result of the Eurocentrism of asymmetrical concepts: at the time when these concepts were coined, noone could imagine a region West of Europe.[10] But this also clarifies the unique status and the revolutionary impact of the American discoveries for the self-understanding of Europe. Although it was pretty straightforward to interpret the encounter with the New World in terms of the traditional distinctions 'civilization' vs. 'barbarism' and 'Christians' vs. 'pagans', it was of utmost importance that "the emerging new world did not appear as a new enemy but as an open space, a free zone of European occupation and expansion" (Schmitt 1997: 55). The colonial discourse was less about danger and threat then about supremacy and domination. And as the Europeans themselves were now in the position of 'Eastern invaders', they could perhaps better identify themselves with the Eastern component in the conceptual asymmetries that had structured the European discourse. The way was open to acknowledge one's own barbarian past.

(2) In his original study Reinhart Koselleck emphasized the possibility of reviving the old asymmetrical concepts within the later ones (Koselleck 1989: 217). The late 16th and early 17th century is a time in which the religious contrast between Christians and heathens lost its strength. The colonial competition between the old Catholic nations Spain and Portugal and the new Protestant nations England and Netherlands established a threefold scheme: 'European true Christians' *vs.* 'European false Christians' *vs.* 'American heathens', and in this new moral hierarchy the heathens were located above the heretics (Koselleck 1989). The religious polemic in the context of colonial competition revived the other, older concept: now Protestants claimed that the Spaniards behaved more barbarically than the American savages themselves. The so called 'black legend', the debate about the Spaniards' colonial brutality, can be understood as the re-entry of one asymmetrical concept on one pole of another, structurally similar concept (see above). As we have seen, it was the religious schism which allowed for this internal differentiation of the 'we' in the contrast between us

ferior, subhuman' Russian population in the occupied lands (Gerlach 2001).

10 In the case of the Untermensch-concept we should probably interpret the East-West divide as a cultural legacy of the older asymmetrical concepts on which it is based.

Christians and the heathens. In this context it seems worth noting that Luhmann associates re-entry with rationality: the latter exists only insofar as a system, based on the differentiation system vs. environment, uses this difference its own operation, including self-observation, self-awareness etc. (Luhmann 1987: 617, 640). In our context, self-awareness increased in part because the temporality of the newer concept—heathens as would-be Christians—created a new temporal frame for an older concept: before, Europeans too had been savages. Thus asymmetrical concepts were temporalized, and by this token their secular connotations prevailed over the religious ones that dominated before. No longer confined to merely indicating historical change, asymmetrical concepts were now its agents.

All this was of relevance for a new understanding of historical time. De Bry gives

> "evidence of a growing intellectual trend when he appends several portraits of the Picts, inhabitants of ancient Britain, to the account of the settlement of Virginia. Universal decline, whether dated in the classical world from a Golden Age or in the Christian from the fall of Adam and the expulsion from Paradise, had been the leitmotif of Western visions of history. The tendency of Europeans to equate the alleged barbarism of the American populations they encountered with the cultural life of their own ancestors would eventually lead some people to believe that human societies, rather than inevitably deteriorating, progressed through increasingly sophisticated stages of civilization" (Grafton 1992: 129).

By 1800 it had become an accepted world-view: "who sails to the remotest places on earth, is in fact moving on a time line. He is traveling back into the past. With each step, he is led into an earlier century. The islands he reaches are the cradles of human kind" (Joseph-Marie Degerando, quoted in Padgen 1996: 176). But the ground for this understanding had been laid earlier. Apparently, the new idea of historical progress was not a brainchild of the late 18th century bourgeoisie, but rather resulted from dynamization, or temporalization, of asymmetrical concepts in the 17th century.

REFERENCES

Birch, S. 2009. Through an Artist's Eye: Observations on Aspects of Copying in two Groups of Work by John White c. 1585-1590. In *European Visions, American Voices*, edited by K. Sloan. London: The British Museum Press, 85-96.

Boehm, G. 1969. *Studien zur Perspektivität. Philosophie und Kunst in der Frühen Neuzeit*. Heidelberg: Winter.

Bredekamp, H. 1999. *Thomas Hobbes visuelle Strategien. Der Leviathan. Urbild des modernen Staates; Werkillustrationen und Porträts*. Berlin: Akademie Verlag.

Burghartz, S. (editor). 2004. *Inszenierte Welten. Die west- und ostindischen Reisen der Verleger de Bry 1590-1630*. Basel: Schwabe.

Collins, J. R. 2005. *The Allegiance of Thomas Hobbes*. Oxford: Oxford University Press.

Collinson, P. 1988. The Protestant Nation. In *The Birthpangs of Protestant England*, edited by P. Collinson. New York: St. Martin's Press, 1-27.

Corbett, M./Lightbown, R. 1979. *The Comely Frontispiece. The Emblematic Title-page in England 1550-1660*. London: Routledge & Kegan.

Damler, D. 2008. *Wildes Recht. zur Pathogenese des Effektivitätsprinzips in der neuzeitlichen Eigentumslehre*. Berlin: Duncker & Humblot.

Duchet, M./Defert, D./Lestringant, F./Forges, J. (editors). 1987. *L'Amerique de Théodore de Bry. Une Collection de voyages protestante de XVIe siècle. Quatres études d'iconographie*. Paris: Editions du Centre National de la Recherche Scientifique.

Feest, C. F. 2007. John White's New World. In *A New World. England's first View of America*, edited by K. Sloan. London: The British Museum Press, 65-77.

Gaudio, M. 2008. *Engraving the Savage. The New World and Techniques of Civilization*. Minneapolis: University of Minnesota Press.

Gerlach, C. 2001. *Krieg, Ernährung, Völkermord. Deutsche Vernichtungspolitik im Zweiten Weltkrieg*. Hamburg: Pendo.

Grafton, A. 1992. *New Worlds, Ancient Texts. The Power of Tradition and the Shock of Discovery*. Cambridge, Mass.: The Belknap Press of Harvard University Press.

Greve, A. 2004. *Die Konstruktion Amerikas. Bilderpolitik in den Grands Voyages aus der Werkstatt de Bry*. Köln: Böhlau.

Harriot, Th. 1972. *A Briefe and True Report of the New Found Land of Virginia. The Complete 1590 Engravings by Theodor de Bry after the Drawings of John White and other Illustrations*. New York: Dover Publications.

Hobbes, Th. [1651] 1996. *Leviathan*. (edited with an Introduction and Notes by J. C. A. Gaskin). Oxford: Oxford University Press.

Hobbes, Th. [1650] 2005. *The Elements of Law, natural and politic*. London: Elibron.

Hobbes, Th. [1642] 2007. *On the Citizen*. (edited and translated by R. Tuck/M. Silverthorne). Cambridge: Cambridge University Press.

Hoekstra, K. 2007. Hobbes on the Natural Condition of Mankind. In *The Cambridge Compendium to Hobbes's Leviathan*, edited by P. Springborg. Cambridge: Cambridge University Press, 109-127.

Hodgen, M. T. 1964. *Early Anthropology in the Sixteenth and Seventeenth Century*. Philadelphia: University of Pennsylvania Press.

Hulton, P. 1984. *America 1585. The complete drawings of John White*. London and Chapel Hill: The University of North Carolina Press and British Museum Publications.

Hulton, P./Quinn, D. B. (editors). 1964. *The American Drawings of John White, 1577-1590* London and Chapel Hill: Trustees of the British Museum and University of North Carolina Press.

Kendrick, T. D. 1950. *British Antiquity*. London: Methuen.

Kidd, C. 1999. *British Identities before Nationalism: Ethnicity and Nationhood in the Atlantic World, 1600-1800*. New York: Cambridge University Press.

Koselleck, R. [1954] 1973. *Kritik und Krise. Eine Studie zur Pathogenese der bürgerlichen Welt*. Frankfurt am Main: Suhrkamp.

Koselleck, R. [1975] 1989. Zur historisch-politischen Semantik asymmetrischer Gegenbegriffe. In *Vergangene Zukunft. Zur Semantik geschichtlicher Zeiten*, edited by R. Koselleck. Frankfurt am Main: Suhrkamp, 211-259.

Kuhlemann, U. 2007. Between Reproduction, Invention and Propaganda: Theodor de Bry's Engravings after John White's

Watercolours. In *A New World. England's first View of America*, edited by K. Sloan. London: The British Museum Press, 79-92.

Laak, D. van. 2002. *Gespräche in der Sicherheit des Schweigens. Carl Schmitt in der politischen Geistesgeschichte der frühen Bundesrepublik*. Berlin: Akademie.

Lantucci, S. 1972. *I filosofi e i selvaggi, 1580-1780*. Bari: Laterza.

Levy, F. J. 1967. *Tudor historical thought*. San Marino, California: Huntington Press.

Locke, J. [1690] 1988. *Two Treatise of Government*. Cambridge: Cambridge University Press.

Luhmann, N. 1987. *Soziale Systeme. Grundriß einer allgemeinen Theorie*. Frankfurt am Main: Suhrkamp.

Malcolm, N. 2002. Hobbes, Sandys, and the Virginia Company. In *Aspects of Hobbes*, edited by N. Malcolm. Oxford: Oxford University Press, 53-79.

Manow, P. 2011. *Politische Ursprungsphantasien. Der Leviathan und sein Erbe*. München: Konstanz University Press.

Padgen, A. 1996. *Das erfundene Amerika. Der Aufbruch des europäischen Denkens in die Neue Welt*. München: Diederichs.

Pratt, S. 2009. Truth and Artifice in the Visualization of Native People: From the time of John White to the Beginning of the 18th Century. In *European Visions, American Voices*, edited by K. Sloan. London: The Trustees of the British Museum, 33-40.

Quinn, D. B. 1979. *English Plans for North America, The Roankoke Voyages, New England Ventures. Volume III of New American World, A Documentary History of North America to 1612*. London: MacMillan Press.

Scherneck, H. 1996. Außenpolitik, Konfession und nationale Identitätsbildung in der Pamphletistik des elisabethianischen England. In *Nationales Bewußtsein und kollektive Identität. Studien zur Entwicklung des kollektiven Bewußtseins in der Neuzeit 2*, edited by H. Berding. Frankfurt am Main: Suhrkamp, 282-300.

Schmitt, C. [1950] 1997. *Der Nomos der Erde im Völkerrecht des Jus Publicum Europaeum*. Berlin: Duncker & Humblot.

Shields, D. S. 2007. The Genius of Ancient Britian. In *The Atlantic World and Virginia, 1550-1624*, edited by P. C. Mancall. North Carolina: University of North Carolina Press, 489-512.

Skinner, Q. 2008. *Hobbes and Republican Liberty*. Cambridge: Cambridge University Press.

Sloan, K. (editor). 2007. *A New World. England's first View of America*. London: The British Museum Press.

Sloan, K. (editor). 2009. *European Visions, American Voices*. London: The British Museum Press.

Smiles, S. 2009. John White and British Antiquity. Savage Origins in the Context of Tudor Historiography. In *European Visions, American Voices*, edited by K. Sloan. London: The British Museum Press, 106-112.

Tuck, R. 1996. Introduction. In *Thomas Hobbes, Leviathan*, edited by R. Tuck. Cambridge: Cambridge University Press, IX-XLV.

Tuck, R. 1998. Introduction. In *Thomas Hobbes, On the Citizen* edited by R. Tuck/M. Silverthorne. Cambridge: Cambridge University Press, VIII-XXXVII.

Uhlig, C. 1996. Europäische Nationalepik der Renaissance. Fortschreibung und Korrektur der Antike. In *Nationales Bewußtsein und kollektive Identität. Studien zur Entwicklung des kollektiven Bewußtseins in der Neuzeit 2*, edited by H. Berding. Frankfurt am Main: Suhrkamp, 249-281.

On Histories, Revolutions, and the Masses

Visions of Asymmetry and Symmetry in German Social Sciences

JAN MARCO SAWILLA

INTRODUCTION

For a long time, the discussion of asymmetry seemed identical with the discussion of chaos (McManus 2005: 157-162). In his 1949 study "On the Problem of Symmetry in Art", the Austrian art historian Dagobert Frey (1883–1962) described some of the main attributes associated with symmetrical and asymmetrical structures in terms of a binary semantic field: "[...] the order of symmetry corresponds [...] to the arbitrariness of asymmetry, the lawfulness to the accident" and "the binding to the freedom, the petrification to the life". Symmetry was "rooted in formal rigidity and constraint"; asymmetry, in "playfulness [...] or changeability and ambiguity" (Frey 1949: 276). In the same issue of the periodical *Studium Generale*, authors including the physician and chemist Karl Lothar Wolf (1901–1969), the botanist Wilhelm Troll (1897–1978), and the mineralogist Wolf von Engelhardt (1910–2008) discussed the question of symmetry as essentially one of concordant and well-proportioned forms. Wolf remarked that one could speak of symmetry if a "(beautiful) relation between one part and another and between the parts and the whole" existed (Wolf 1949: 213). Troll attempted to demonstrate that the beauty of symmetrical and well-proportioned arrangements—even in botany—could be explained by interpreting them as a synthesis of natural primordial form (*Urbild*) and of expediency (*Zweckmäßigkeit*); a synthesis that would be responsible for the development of the shape of natural entities in general (Troll 1949: 259-264). According to von Engelhardt, human beings would be unable to orient themselves in their environment without possessing the capacity "to see symmetries everywhere" (Engelhardt 1949: 203).

These approaches focused on the comparison between natural objects and the arts, which were understood to be the most sophisticated expression of the modes of operation of the human mind, and the theories represented the general state of research at this time. These theories were later adapted by the physician and philosopher Hermann Weyl (1885-1955). In his widely known book *Symmetry* (1952), Weyl sought to explain "where the precise geometric notion of bilateral symmetry begins to dissolve into the vague notion of *Ausgewogenheit*, balanced design" (Weyl 1952: 16).[1] From his point of view, symmetry

1 While the German Weyl was a scientist in exile, Wolf and Frey were active as political scientists in Germany at the time of the Third Reich. As president of the University of Kiel, between 1933 and 1935 Wolf fostered the reorganisation of the University under the Nazi regime. Later, as a professor in Halle, he did research in service of the German war industry (Grüttner 2004: 185–186; and URL: http://www.catalogus-professorum-halensis.de/wolfkarllothar.html [01.04.2011]). Nevertheless, he was a supporter of a holistic approach directed against a fundamentalist and purely racist understanding of Darwinism and other branches of contemporary biology. Wolf kept in contact with the circle of Stefan George (1868–1933). Together with Troll and the art historian Wilhelm Pinder (1878–1947), he published the journal *Die Gestalt. Abhandlungen zu einer allgemeinen Morphologie*—including publication of his book *Gestalt und Symmetrie* (1952)—and beginning in 1941 was a founding member of the editorial board of the *Leopoldina Ausgabe*, in which the writings of Goethe on nature and natural science were edited (Breuer 2005: 306-307; and: Goethe. Die Schriften zur Naturwissenschaft [Leopoldina-Ausgabe]. Geschichte der Edition. http://www.leopoldina.org/goethe/gesch.htm [01.04.2011]). Engelhardt was a member of SA between 1933 and 1934 and in 1939. Between 1942 and 1945, he was in service of the SD. In addition to the *Nationalsozialistischer Deutscher Dozentenbund*, he was a member of other

was inherently linked to questions of beauty and order (McManus 2005: 157-160). The main heuristic tool to identify symmetrical forms was the development of ideal types. As Wolf and the German studies specialist Dorothea Kuhn describe, as a consequence of "the difference between idea and figure the symmetry of objects never becomes visible in an entirely pure manner". Very often, it even appears in a "cloaked" form (Wolf and Kuhn 1952: 35).

Within scientific historiography, "symmetry" and "asymmetry" are concepts of little significance. There is probably no introduction of the study of history that mentions either of these concepts. However, the notion of ideal types is well established (Rohlfes 2005: 74-77). Likewise, the terms associated with notions of symmetry and asymmetry mentioned above played and still play an important role in the analysis and evaluation of historical processes. For example, in his noted book *Crisis of the European Mind*, the French historian Paul Hazard (1878-1944) characterised the period between 1680 and 1715 as an era of "crisis"; to him, this meant an era of "dissonance", metaphorically speaking. Neither the "harmonies" of "lyrical" sounds nor the "miraculous equilibrium" (*équilibre miraculeux*) of the preceding Classical Age, but rather the aggressive appearance of the new and the essential disturbance of traditional ways of creating social stability became the hallmarks of the late seventeenth century. Old connectors either proved useless or lost their appeal almost overnight *(tout d'un coup)*, while new ones had barely been established (Hazard 1961: 10, 15, 38). Without a theoretical framework at his disposal to organise his obser-

National Socialist organizations. Until the end of World War II, he was a lecturer and professor at the University of Göttingen, and from 1957 to 1978 a professor at the University of Tübingen. In 1970, he also became one of the editors of the *Leopoldina* (Szabó 2000: 308; Wyder: 2008). Frey, a professor at the University of Breslau since 1931, was involved in the cultural and historical appropriation of the eastern territories after the German assault on Poland (Störtkuhl: 2007; Fuhrmeister 2008: 326, 330-331; Kott 2008: 366n.; Papenbrock 2008: 26-27; not mentioned in the biographical sketch of Gensbauer-Bendler 1989: 55-59). This is not the place to discuss the possible connections between the aesthetic, scientific, and political ideas of these authors in regard to the problem of symmetry. Suffice it to say that this is no innocuous subject.

vations,[2] Hazard sought to understand and to describe what happens when the epistemological centres of a given society are set in motion.[3]

Hazard's interpretations are not eccentric. Even though the question of beauty as the aesthetic expression of symmetry has never been of overwhelming interest in historiographical contexts, qualities such as order, balance, and if not "harmony" at least the absence of major confrontations seemed to most historians to be crucial for a society's stability. Phenomena of imbalance and the impression of a (perhaps accelerating) loosening of established connections between parts of society and the whole appear to signal the dominance of asymmetrical conditions at a given time, and thus might be paired with a loss of functionality within and in between social systems. Societies described in such terms tend to be identified as being in a state of "crisis", most commonly interpreted as an extraordinary and transient phase located chronologically between two eras of relative stability (Sawilla 2011).[4] To be sure, it is not necessary to categorize phenomena of asymmetry not only as the mere absence of symmetry, but also as an intentional negation of symmetry and order, as Frey did (Frey 1949: 276). However, it is true that almost nobody would describe a society in terms of dissolution or of increasing or intolerably high tension, and would then go on to classify this state as normal or ideal. The notion of normativity seems to go hand in hand with qualities of symmetry.

This essay will focus on two problems connected with the different ways of thinking about and evaluating asymmetry. The first concerns one particular aspect in the development of Reinhart Koselleck's notion of asymmetric counter-concepts. Koselleck's theories are one of the few attempts to link the question of asymmetry to an analysis of

---

2 Consider the theory of "emergence" developed in the first decades of the twentieth century (Stephan 1999: 3–13).

3 He called his opening chapter "From persistency to movement" (Hazard 1961: 15).

4 These periods seem to indicate historical change. For this reason, they are of major scientific relevance and attract a great deal of scholarly attention. This again might be the reason why historians presumably have identified more "crises" in world history than periods of stability. The problem of creating rather than analysing historical "crises" is one of the primary subjects in current debates over the concept of "crisis" (Elliott 2005; Föllmer/Graf 2005; Grunwald/Pfister 2007; Parker 2008; Hölkeskamp 2009).

the formation (and not the decomposition) of social structures. His approach focused on the differences inherent in the description of one's self and the other within a given linguistic system and on the resulting consequences. Koselleck defined as "asymmetric counterconcepts" any "binary concepts" that claim to be true for all members of mankind. They stabilise the identity of social groups by producing a divide between "we" and "they". As "asymmetrical", they can be qualified due to their function as "conflicting labels, employed only in one direction and in an unequal fashion" (Koselleck 2004c: 155–157). In the first part of the essay, there will be a discussion of a possible refinement of this approach, since it is not always clear to what extent the examples given by Koselleck and those who have adapted his reflections must necessarily be considered "asymmetrical" in the proper sense of the word. In addition, Koselleck's use of "asymmetrical counter-concepts" will be contrasted with an earlier usage of the term "counter-concept" in his writings, one that was not a label for "symmetrical" counter-concepts but belonged to a different analytical level.

In the second part, I will pose the question of how to deal with the tool of "asymmetrical counter-concepts" in light of developments in the nineteenth and twentieth centuries, in which societies become hallmarked by a high degree of complexity and a plethora of possible viewpoints and opportunities for social self-description. Clearly, the juxtaposition of "Hellenes" and "Barbarians" posed by Koselleck as a prototype of the analysis of asymmetrical counter-concepts should be applied to more detailed conditions if it is to be adequately evaluated. I will focus on one category that has been studied intensively during the last decade and that is characterised by highly ambiguous semantics and its tendency to be a combat term in both analytical and everyday political and social language: the category of the "masses" (Sloterdijk 2000; Günzel 2004; 2005a, 2005b; Gamper 2007; Lüdemann and Hebekus 2010). Focusing on the scientific debates on the "mass media" that emerged in the 1960s and 1970s, for example, it will be examined whether the "masses" appeared as an asymmetrical counter-concept to the "bourgeoisie" in the writings of those who understood their scientific analyses to be a distinct part of contemporary political discussions. Connecting form with content, it must also be examined whether an entity such as the "masses", most often perceived as amorphous and as anything but well-balanced, necessarily needs a counter-concept in a linguistic system. Under what conditions does the dual structure of asymmetrical counter-concepts begin to lose its value, especially when an entity is observed that itself thinks in terms more advanced than simple oppositions transgresses for itself thinking in terms of simple oppositions? This problem will be investigated with regard to scientific and contemporary descriptions of religious "masses" in the nineteenth and twentieth centuries.

I On Counter-Concepts, Attributes and Relations: Reinhart Koselleck's Reflections on the Roles of "History" and "Revolution" in the Eighteenth Century

Within intellectual history (*Geistesgeschichte*), the term "counter-concept" was already fairly established when Koselleck began his studies in *Begriffsgeschichte*. With the objective of understanding the specifics of a given intellectual and linguistic system, scholars from various disciplines had used this terminology to explain historical expressions that at first glance would not have been considered to be related. For example, the Church historian Karl Wölfl discussed in 1960 *oeconomia/ dispositio* as a counter-concept of *monarchia* and *persona* as a counter-concept of *substantia* in Tertullian (Wölfl 1960: 74-75n, 100). Herbert Marcuse brought to light the notion of *Ansichsein* as the "Gegenbegriff" of *Dasein* in the philosophy of Hegel (Marcuse 1968: 59). Joseph H. Kaiser explained that *Bürgertum* was the "Gegenbegriff" of *absolutistischer Staat* in political language from the first half of the nineteenth century (Kaiser 1949: 124). Koselleck used the term "counter-concept" in a similar way. The most familiar example of this is his interpretation of the terms "history" (*Geschichte*) and "revolution" (*Revolution*) and their function within in the emerging High German language at the end of the eighteenth century.

The concepts of "revolution" and "history" occupy a paradigmatic position within Koselleck's semantics of historical times. According to his analysis, both terms became collective singular nouns in the last decades of the eighteenth century. This means that they became polysemantic and abstract concepts—i.e., concepts in the proper sense of the word.[5] Up to that point, the nouns "history" and "revolution" had been countable. They were used either to refer to single cases or to the sum of these cases, predominantly in plural forms (Koselleck 1975: 594, 653-657, 711; Koselleck 1984: 732-739; Koselleck 1988; Koselleck 2001: 18-21; Koselleck 2004a: 35). It is not necessary to discuss here in detail the difficulties associated with the notion of the process that Koselleck calls "singularization". These difficulties are essentially of empirical and semantic nature. They result on the one hand from the fact that at least in French—and presumably also in English—from about the 1670s on, a concept of "history" (*histoire*) existed, characterised by the polysemantic structure considered by Koselleck to be a hallmark of the German term *Geschichte* one century later: that is, "history" as a word that describes the totality of historical events, the historiographical tradition, and a specific technique dealing with these events and traditions. On the other hand, it should be noted that prior to the last decades of the eighteenth century, the early modern German term *Geschichten* was not a simple countable noun signifying the sum of single "histories", but rather was a transnumeral modelled on the Latin pluraletantum *res gestae*. This linguistic characteristic was specific to the early modern German language (Sawilla 2004; Sawilla 2010: 396-408, 420).[6]

More important for the present essay is the question of why "history" and "revolution" can be considered counter-concepts. Koselleck's answer was twofold in nature. First, both "history" and "revolution" not only became collective nouns at the end of the eighteenth century but also meta-historical concepts as well. They were thought to be ap-

5 For the theoretical background, see especially Bödeker 2002: 85-90; Wiehl 2003: 87-88; Sawilla 2004: 414-416. An overview of the literature is provided by Gumbrecht 2006; Steinmetz 2008; Burke 2010, and the contributions in the volume of Joas/Vogt 2010.

6 It must be clarified whether it is still helpful to stress not only the formal but the semantic implications of the process of "singularization" as well; see the summary of Burke (2010: 153-154). Since the attempt to connect the history of the word with the history of the concept (one of the main methodological features of German *Begriffsgeschichte*) and to distinguish it from similar approaches such as the History of Ideas or the History of Concepts (Richter 1995) also causes many problems, the particular capacity of *Begriffsgeschichte* must be examined as well.

plicable to all members of mankind and, as universal categories, they reached the level of basic principles regulating the perception of a changing world. Second, while "history" (in association with the emergence of the concept of "progress") was developing into a dynamic force in its own right, the concept of "revolution" helped to evoke and describe the experience of an accelerating time: an experience that led to the possibilities of a radical break with the past, of considering an essentially unknown future, and of understanding the future as the main objective of actions in the present. In this sense, according to Koselleck, the French Revolution as an historical event was prepared by the new experience of genuine historical time—by the notion of the possibility of "making history"—and its reflection over the course of the eighteenth century. On the other hand, the perceived uniqueness of the French Revolution promoted and stabilised the (modern) experience of the mutability of all things. Koselleck himself described the "mutual relationship" between the concepts of "history" and "revolution". Not considering this relationship would mean misunderstanding the characteristics of the modern concept of "revolution" (Koselleck 1984: 734-739; Koselleck 2004a: 35-42; Koselleck 2004b).

Koselleck's use of the notion of "counter-concept" in this situation is not a systematic one. The relationship between "history" and "revolution" does not correspond to a relationship between a whole and its parts or between the particular and the general. Neither does it correspond to a difference between the description of one's self and the other– on that score, with regard to the period following the French Revolution, Koselleck denoted "evolution" and "reform" and later "reaction" and "restoration" as asymmetrical counter-concepts of "revolution" (Koselleck 1984: 749-760). Rather, the relationship between "history" and "revolution" is defined in terms of functional complementarity and semantic similarity (Koselleck 1984: 720-721, 739-740) concerning their crucial roles in organising an essentially new area of historical experience and its linguistic expression.[7] Generally speaking, apart from a few further variations,[8] Koselleck's use of "counter-

7 A further example for the assumption of functional complementarity is the assumption that "both [concepts] smashed the preceding models, while at the same time apparently incorporating them" (Koselleck 2004a: 35).

8 Sometimes Koselleck qualified "revolution" as a "form of expression of 'history in general'" ("Erscheinungsform der 'Geschichte überhaupt'")

concepts" was an important stage in overcoming the author-focused methodology of the prior *Geistesgeschichte* and in approaching step-by-step the field of analysis of semantics (Bär 2000). Discussing "counter-concepts" in such a way, however, does not imply the oppositional momentum constitutive of asymmetrical counter-concepts. For that reason, one should perhaps designate them as "congruent counter-concepts." The identification of these counter-concepts is the result of analytical operations. It does not depend greatly on an examination of the way historical societies describe themselves and their perceived "others", but rather emerges from a search for structural similarities and congruencies. Further research will be needed to clarify how the nature of this relationship could be more precisely defined. However, this would mean shifting the focus from the question of attribution, from the establishment of sets of binary and contrary qualities associated with particular terms or objects, to that of the social and linguistic construction of the relationship itself. This seems to be all the more interesting because in current philosophical and sociological theories, symmetry and asymmetry are primarily discussed as a problem of relationships and their dynamics.[9] Such a shift would enable more accu-

(Koselleck 1984: 740). In this case, the problem of counter-concepts is not subject to debate. In addition, Koselleck seems to have occasionally advanced the view that the French word *révolution* could be regarded as a kind of counter-concept of the German *Geschichte*, since while the French were transposing the concept of "revolution" into reality, since 1789 the Germans had thought of "history" in the quality of a historical force in its own right (Dipper 2000: 300). In that case, one must talk about the functional equivalence of different concepts in two distinct languages. However, one should not overemphasize this aspect. Koselleck knew, of course, that the formula of *philosophie de l'histoire*—which is the focal point of his interpretation—was used for the first time by Voltaire (Koselleck 2004a: 36 and n. 57). This means that the French for their part were also thinking about "history", and that they did so before the Revolution and also at the time of the Revolution (see in detail Sawilla 2010: 409-414).

9 This is the initial point of an approach that aims to explain why the process of creating and establishing asymmetries ("Asymmetrization") should be regarded not as a disturbance, but rather as one of the "basic requirements of the constitution" (Krause 2001: 107) of social systems. According to Niklas Luhmann, this process is one of self-selectivity. It concerns the ca-

rate recognition of the way asymmetrical counter-concepts work as well. A few years ago, Kari Palonen posed the question of whether the counter-concepts "friend" and "enemy" should be indeed described as symmetrical (as proposed by Koselleck) or as asymmetrical, as Carl Schmitt claims (Palonen 2004: 225). What is the difference between asymmetrical and symmetrical counter-concepts?

According to the approach of formal logic, a "binary relation R is called symmetric if xRy implies yRx". Unlike this "mirror-symmetry", a relation is denoted as non-symmetric or asymmetric if the elements involved do not coincide "with [their] converse"; for example, the relations "'being heavier than' or 'being colder than'" (Lorenz 2005: 3, 5). With this in mind, it is obvious that a relation "Hellenes" – "Barbarians" cannot be qualified as symmetrical or asymmetrical in a general way. It could be called symmetrical if the relation were "being different from"; in this case, it would be a typical "mirror-symmetry". It must be called asymmetrical, however, if the relation implies a comparison such as "being less civilised than"—assuming that the "Barbarians" do not perceive the "Hellenes" as "less civilised", for then one has to discuss symmetry once more. It is possible to think of even more intricate examples and in this way approach the complexity of social reality. The relation "Christians" – "Polytheists" described as "are venerating a god as well as" is both symmetrical and asymmetrical, depending on the perspective. From a neutral point of view, it must be called symmetrical and asymmetrical when taking into consideration the respective, mutually exclusive angle, for it is established

---

pacity of a vital social system to be both connected to its environment and self-contained in building and maintaining its structures: "[...] a complex system [...] cannot rely on point-for-point correspondences with the environment. It must *give up the idea of full synchronization with the environment* [...]." (Luhmann 1995: 37-39, 43) Concerning its internal construction—the shaping of subsystems—as well as the relation between environment and system and including aspects of diachrony as well as of synchrony, asymmetrical relations are essentially responsible for the differentiation of complexity. Whatever one might say in detail about this process of "asymmetrization", it always gives rise to something. It is "rich in consequences" (Luhmann 1995: 192) and cannot be described as a chronological succession with the meaning of: "from a state of symmetry to a state of asymmetry and possibly back to a state of symmetry".

that the Romans were able to recognise the Christian god as a divinity, while the Christians were convinced that the Romans venerated only idols (Assmann 2008). With his notion of "asymmetrically opposed concepts" (Koselleck 2004c: 156), Koselleck emphasised this relation in particular. The same seems to be true for the historiographical adaption of the tool, even when the objects and enquiries became more sophisticated, as with the analysis of processes of cultural translation in its linguistic, social, and political implications, taking into account the respective views of the scientists themselves (e.g., Heinke 2005: 100-101; Laak 2005: 18; Osterhammel 2006: 276; Shimada 2007: 207; Wiedemann 2007: 28-29; Schröder 2009: 237, 311). In the following pages, it will be asked to what extent the study of asymmetrical counter-concepts could be modified by factoring in the question of attribution as well as that of creating relationships and describing social phenomenons from different points of view.

II The Bourgeoisie, the Marxists, and the "Masses": On Social Progress and Its Opponents

In his essay *Crowds and Power*, published in 1960, Elias Canetti (1905-1994) connected the appearance of the "masses" in a given situation to the impression of abruptness: "A few people have been standing together—five, ten or twelve, not more; nothing has been announced, nothing is expected. Suddenly, everywhere is black with people [*schwarz von Menschen*] and more come streaming from all sides as though streets had only one direction" (Canetti 1960: 14). Since the times of Greek and Roman antiquity, the "crowd" and the "masses" have been depicted as an object of seduction, associated with qualities such as incalculability, fickleness, and the tendency to be the opposite of social order (Gamper 2007: 13-15). The transition from monarchic to (in a broad sense) democratic systems by the end of the eighteenth century and especially the events of the French Revolution generated an idea of the "masses" as an entity acting not so much in its own right (Lüdemann/Hebekus 2010: 8),[10] but rather as an multiply valued unit. Simultaneously with other processes such as industrialisation and its impact on human coexistence, the "masses" changed their structural position within the description of societies. They were correlated with other categories of growing, modified, or even new significance. Regarded as a social phenomenon of "great number" *par excellence* they could be understood as a temporarily transient and immediately perceivable force. In this capacity, they were seen as a counterpart to more abstract and durable units such as "population" and "assembly", and "nation" and "class", the latter two considered to be the first beneficiaries of the possible redistribution of political power (Gamper 2007: 19-29; Lüdemann/Hebekus 2010: 10).

Of course, the "masses" could still be described in terms of absent qualities, such as presumed lack of structure, rationality, durability, or civility. Especially due to their energetic and dynamic characteristics, however, the "crowd" and "the masses" began to appear as an appreciably controversial category as well. These terms evolved a combat term attracting reflections about the movements of society as a whole. In Marxist theory, the "masses" found their way out of their previously amorphous existence by becoming aware of their own status, developing a consciousness of themselves as a "class", and thus emerging as an essential font of human progress in history in general (Schönemann 1978: 366-368; Günzel 2004: 121-126). Psychophysiological approaches have outlined the actions of the "masses" as the nucleus of vivid statehood, cyclically regenerating its energies and unconsciously advancing social progression (Wurst 2010).

Briefly stated, within the political, social, and scientific language of the nineteenth and twentieth centuries, the category "masses" evolved as a relational and highly polarising concept that simultaneously evoked notions of positivity and negativity and tendencies of identification and alienation, depending on one's perspective. One of

---

10 During the so-called "Fettmilch Uprising", for example, the council of the city of Frankfurt complained in an memorial of August 1612 about the pressure created by the threat posed by the "uncontrollable furiousness" of "several hundreds, yes even thousands of people" on the street and the ungovernable "storminess" (*Ungestüm*) of this "crowd" (*Menge*) (*Ungefährlicher Verfaß, welchergestalt Ein Ehrbarer Rat Narrata Mandati zu verificieren hat*, Bothe 1920: 368-381, 372).

the main divides ran between the "masses" and the "bourgeoisie". This dialectic division could be compared to the construction of the asymmetrical counter-concepts "Hellenes" and "Barbarians" as described by Koselleck, essentially organised by a "we" and "they" binary pair. However, as evaluated by Koselleck's criteria, neither terms were universal categories. Is it still possible to speak of asymmetrical counter-concepts in terms of the "masses" and the "bourgeoisie"? Concerning the question of relationality, it is indeed possible to establish some direct and reciprocally valid relations between these units, such as "inhibit social progress".[11] However, other relations that are part of the contemporary set of attributions applied to the respective "other" would make no sense at all if one tried to integrate them into a relational framework characterised by reciprocity. The proposition "are a threat to social stability" would have been useless for a Marxism-based understanding of the role of the "bourgeoisie", while this perspective would have had no difficulties in describing the Marxist "masses" in such terms. Therefore, the tool "asymmetric counter-concept" as defined by Koselleck might be a starting point to analyse semantic fields or discursive formations, but it seems to be conceptually limited when considering in detail ways of creating social oppositions, especially taking into consideration the plurality and heterogeneity of standpoints within a modern socio-political framework, the number of potential documents to examine, and the necessity to bear in mind the historical position of the modern scientists themselves. These problems can be illustrated by outlining two different ways of handling the "masses" in historical and social analysis. The first can be observed in Jürgen Habermas' book *The Structural Transformation of the Public Sphere* (1962) and in the early theories of mass media that emerged around 1970. The second addresses the thesis that aside from the ambiguity characterising the "masses" on their own, the latter tend to undermine the possibility of fitting into a dualistic system; this becomes apparent when the bourgeoisie and other observers abandon their analytical points of view and try to depict what they see when they mix themselves with the "masses".

11 This relation should be called symmetrical for the reasons mentioned above.

II.I "Knowledge without Consciousness": Theories on Mass Media

The "masses" represent not only a social unit but also seem to denote a particular experience of modern existence in general. Inasmuch as the category in most cases reflects experiences and phenomena perceived as negative—mass unemployment, massive groups of livestock, or even weapons of mass destruction or mass graves—it is not just an expression of intellectual elitism and "contempt for the masses," as Peter Sloterdijk recently stressed, but it is also commonly accompanied by the pretension of critical social analysis as well. This is the case in Habermas' *Structural Transformation of the Public Sphere*, first published in 1962. In scientific historiography, this study has been received as a delineation of the transition from the "representative publicness" of the seventeenth century to the "bourgeois public sphere" in the course of the eighteenth century (Schiewe 2003: 253-266; Rau/ Schwerhoff 2004; Schlögl 2008). However, Habermas described a second transformation, that from the bourgeois public sphere to the age of mass media. Against the backdrop of twentieth-century totalitarianism, his analysis of the "pathways in the disintegration of the bourgeois public sphere" (Habermas 1989: 175) was characterized by its pessimism.

According to Habermas, since the last decades of the nineteenth century, there has been an observable shift "from a culture-debating [*kulturräsonnierenden*] to a culture-consuming public" (Habermas 1989: 159). The latter is the domain of "the masses". In connection to the falling price of books, the culture of the masses (to mention only a few points) is determined by the psychological facilitation of access to cultural products. In this way the market, adapting the needs for entertainment and relaxation, inverts the prior tendency in which, as Habermas puts it, "the 'people' were brought up to the level of culture" by means of cheap books; now, "culture was [...] lowered to that of the masses" by means of "economically" as well as "psychologically" cheap books. Aside from the "needs of mass taste" and their continuously growing importance, even the books themselves evaded criticism because of their sheer quantity ("great mass") (Habermas 1989: 166-167).

As delineated by Habermas, these and similar shifts are accompanied by the reorganisation of the divides between public and private in general. Due to the technical requirements of traffic flow, the changing infrastructure of the modern city destroys former areas of public communication. In combination with an architectural style that leaves no space for public contact between private people, modern existence tends to be in part an existence of the "masses" on the streets and in part an existence within the intimacy of the nuclear family (Habermas 1989: 154-159). The rational-critical debate among private people articulating the interests of society with the state has lost its space. It has been replaced with new institutions, such as associations (*Verbände*) or parties "aimed at the ideological integration and the political mobilization of the broad voting masses" (Habermas 1989: 196-204, quotation 203). In turn, the mass media not only absorb the elements of the former "culture-debating", subordinating them to the guidelines of entertaining and "human interest", but also assimilate "elements of advertising" and become in that extent essentially apolitical as they foster specific political and economical interests as well (Habermas 1989: 175, 189-194). In short, as Habermas states, "The world fashioned by the mass media is a public sphere in appearance only" (Habermas 1989: 171). The "mass opinion" has almost nothing in common with the former "public opinion" (Habermas 1989: 240).

It should be noted that Habermas revised important parts of his interpretation in a new edition of his book in 1990 (Schicha 2003: 115-116). However, in 1962, unlike the theories of mass communication that followed him, he did not distinguish decisively between the addressees of media products and an understanding of "mass" (within the compound term "mass media") as an entity *enabling* communication of different kinds of messages to a large and potentially enormous number of people (Böckelmann 1975: 35). This is a crucial point, since some theories of mass media have primarily been interested in the technical conditions of the production and distribution of media (Hickethier 2004: 39); the addressees in these theories played a negligible role. Others have focused specifically on the structure of the recipients (Schneider 2002). In this case, the amorphous unit "masses" quickly lost its relevance due to its restricted analytical value.

The same is true, in a different way, of the Marxism-influenced theories of media that evolved in the context of the student movement of 1968. In 1973, for example, Oskar Negt and Alexander Kluge proposed the establishment of a "proletarian public sphere" as an answer to the lack of opportunities (from their perspective) for the political

Left and the members of mass organisations (*Massenorganisationen*) to articulate their interests outside of the "bourgeois public sphere" (Negt/Kluge 1973: 7). While adapting a specific theoretical framework, Negt and Kluge generally relinquished the term "masses" and relied on more precise terminology. To refer to great numbers of people, they spoke of the "working class", "proletarians", or "collectives." However, at the very beginning of their analysis, Negt and Kluge presented a few observations referring to the pragmatic linguistic usage of the KPD (German Communist Party) concerning the question of the "masses". They remarked that the delegates of the KPD referred on their assemblies on the "proletarian public" when controversial decisions should be made by invoking the (alleged) acclaim of those who were politically represented by the party. According to Negt and Kluge, this appeal to the "masses" should be considered as a peculiarity of the "bourgeois public sphere" and designated an advanced state of alienation between the party and the "masses". Thus, in concordance with their theoretical foundations, Negt and Kluge tended to equalise "proletarian public sphere" with the "masses" and, more precisely, with the "party" (Negt/Kluge 1973: 10-11).

Negt, for his part, addressed the question of mass media itself a few years later. In compliance with the modification of the object of analysis and the concomitant relocation of the dominant points of identification, the "masses" appeared in this essay again as an amorphous unit of people who, as consumers, are kept from gaining consciousness of their own social position or from participating (as previous *citoyens* had) in the process of social emancipation. Similar to Habermas' interpretation, Negt saw television in particular as a symptom of an "irretrievable process of decay" of the "bourgeois public sphere" (Negt 1978: 65; Brantlinger 1983: 41-42, 249-250). The "masses" were not only characterised by a lack of self-knowledge—Franz Dröge studied this notion of "knowledge without consciousness" in detail in 1972—but they also had lost their previous energy, once considered as either destructive or progressive, in favour of a largely "apathetic" existence (Dröge 1972: 171).[12] Arising from "the joint efforts of subliminal co-

12 The semantic field of the early theories of mass media was essentially structured around the concepts of mass psychology that had emerged around 1900 and their derivatives, categories including hypnosis and suggestion (Bartz 2007: 105-116).

ercion by mass media and manifest coercion by police" (Negt 1978: 65) these "masses" were creations (or rather, creatures) of the refined methods of authoritarian oppression.

According to Michael Gamper, the relevance of the concept "masses" as an object of scientific interest was in decline by the 1930s. The "semantic doubling of uprising and levelling", which had been characteristic of the perception and description of the "masses" to that point, was replaced by impressions of "conformity" and "normality" that would shape the scientific treatment of the "masses" in the post-war era (Gamper 2007: 511). The analysis of the social role of the mass media and the emotionalised manner of its realisation might not support Gamper's thesis of this loss of relevance. His diagnosis of a shift in the distribution of the attributes associated with the "masses", however, seems to be applicable to this sector as well (see in detail Bartz 2007), even when it is not completely clear whether the "masses" of the mass media largely converge with a more traditional understanding of "masses", meaning the religious "masses". This will be described in the following passages before returning to the question of asymmetrical counter-concepts.

II.II On Binarities and Their Complicating: Shifting Viewpoints on Religious "Masses"

As observed to this point, social analysis seems to become more precise when the use of counter-concepts in social and historical analysis is relinquished.[13] For example, in his sophisticated *German Social History* Hans-Ulrich Wehler had no had no biases regarding the word and the concept of the "masses". In the first volume (published in 1987), he even described the events of the French Revolution almost entirely without reference to the "masses". This is not self-evident, since most writings about the French Revolution in the 1970s discussed at length the importance that could be attributed to the "masses", "the crowd", and "the people on the streets" within the Revolution as a focus of ac-

13 This does not mean that the analysis of the existence of counter-concepts in a given linguistic and social system should not be analysed.

celerated social change.[14] In his reading of the events, Wehler avoided all sorts of binary oppositions, instead describing the events and their causes by making reference to specific professions ("the peasants"), to specific strata ("the impoverished petty bourgeoisie", "the urban bourgeoisie"), to certain classes ("the aristocracy"), and to contemporary parties ("the Jacobins") (Wehler 1987a: 355-362). The same is true of his description of the German Revolution of 1848 (Wehler 1987b: 703-758). As far as can be determined, Wehler relied in only one case on the term "masses" in its traditional sense: namely, when describing the multitudes of religious people at one of the most prominent religious events in nineteenth-century Germany, the Pilgrimage to Trier in 1844. In his explanations, Wehler was supported by a previous essay by Wolfgang Schieder published in 1974. What was the nature of this event, and how was it interpreted by contemporaries and later scientific historiography?

In 1844, Wilhelm Arnoldi (1842-1864), the bishop of Trier, decided to display the *tunica Christi* between August and October in the cathedral of Trier. The tunic had been "recovered" in 1512 from underneath the high altar of the cathedral and it had been displayed several times in the course of the early modern era (Seibrich 1996). It was one of the most important relics owned by the bishopric of Trier and symbolised in its seamlessness the unity of the church. The exposition of the tunic and the pilgrimage to it in 1844 attracted between 500,000 and 1,000,000 people (Schieder 1974: 421-424). Exact figures do not exist; all estimates are based on figures promoted by the bishopric itself. The event provoked an extremely controversial and emotional debate. In Protestant journals, for example, one could read about a

14 For example, it was asked whether the pressure of the "masses"—who most commonly appeared in this context as a part of the formula "the rural and urban masses"—prevented the bourgeoisie from instituting a generally restorative policy. It was asked whether the activities of the "masses" on the streets were more or less important than the constitutional reorganisation of the state or whether the temporary alliance between "the rural and urban masses" and the bourgeoisie was in the end more profitable for one social unit or the other (Furet 1976: 51; Schmitt 1976: 15-18; Soboul 1976: 51; Mazauric 1976: 109). To put it simply, the lines between the different interpretations of the "masses" correspond in this case to the degree of adaption of Marxist or Neo-Marxist models (Knapp/Pelzer 2004).

conspiracy directed by the Jesuit Order with the aim of separating the political province of Rhine from Prussia by means of mobilising the ultramontanist flank of Catholicism in Germany (Scheitler 2000: XVII–XVIII). The essential divide, however, ran not between Catholicism and Protestantism but between ultramontanist Catholicism and so-called German Catholicism (*Deutschkatholizismus*). The main figures on the latter side were the Prussian curate Johann Czerski (1813-1893) and the Silesian chaplain Johannes Ronge (1813-1887), who polemicised against the alleged subordination of German Catholicism to the interests of Rome. Both Czerski and Ronge were excommunicated in 1844 (Schneider 1996a; Schneider 1996b; Steinruck 1996).

In his interpretation of the event, Wolfgang Schieder emphasised that the pilgrimage was a result of the alliance between a group of ultramontanist clerics and aristocratic Catholics, an alliance that used their historical and theological knowledge to manipulate the devotional needs of "the masses". Although it is impossible to be precise about the social composition of the pilgrims, Schieder stressed the fact that almost no members of the learned and wealthy bourgeoisie, almost no people with a university education, and almost no members of the bureaucracy participated. In short, allegedly no sophisticated or emancipated forces took part. For this reason, Schieder evaluated the pilgrimage as a movement of the lower strata. Because of their lack of education, these people were unable to be informed by the public debate over the historical facts and the manipulative nature of the event. According to Schieder, the pilgrimage sent a loud and negative signal that still had effects during the Revolution of 1848: it was an instrument of ultramontanist Catholicism to exercise and to affirm its conservative influence on the "masses" (Schieder 1974: 419, 428-429, 441, 449-451, 454). The church historian Rudolf Lill discussed the main methodological problems of Schieder's approach (Lill 1978) in a reply published in 1978; Wehler, in turn, insisted on the theory of the pilgrimage as an exploitation of a largely unlearned and unemancipated "mass" for political ends that noticeably inhibited the speed of social development (Wehler 1987b: 472-474).

However, it seems obvious that Schieder reproduced to a great extent the arguments and the perspectives of the opponents of the pilgrimage. For example, Ronge discussed the "stifling, ignorant, dull, superstitious and in part degenerated" masses, predominantly rural representatives of the "lower classes", in their pilgrimage to the tunic

(Heil.-Rock-Album 1844: 64). The correspondents of the *Deutsche Allgemeine Zeitung* published in Berlin also observed mainly people from the "lowest classes", most of them women, looking "poor and pitiful" (Heil.-Rock-Album 1844: 48). In contrast, the ultramontanist Catholics emphasised that all social strata participated.[15] The "masses" were in this context the traditional unit of "they", a unit no one was eager to be identified with; however, the pilgrimage was also an embodiment of what Karl Marx (1818–1883) had indicated a few months before in his *Criticism of the Hegelian Philosophy of Right* as the main effect of religion, namely as "the opium of the people" (*Opium des Volks*) (Eßbach 2010: 62).

It would be interesting to broaden these observations by studying contemporary descriptions of people visiting other important pilgrimage locations in the nineteenth century, such as Loreto in Italy, Mariazell in Austria, or Altötting in Germany. For the current context, however, it is sufficient to impart a short impression of Lourdes pilgrimage (Harris 1999; Kaufman 2005; Kotulla 2006), particularly of "Les foules de Lourdes" published by Joris-Karl Huysmans (1848-1907) in 1906. Huysmans' approach, which follows Émile Zola's (1840-1902) experimental attempt to connect the scientific examination of social phenomena to literary forms of expression (Gamper 2007: 372-392), is of particular interest. Huysmans did not use a set of simple dichotomies as Ronge did, in the sense of "we" who are able to recognise the manipulative character of a religious event and "the others" who are manipulated or the manipulators. Rather, he attempted to depict from the standpoint of an observer and distant participant the inner dynamics of the emergence and locomotion of a "mass". He described, for example, several groups of pilgrims arriving at the town as the "indolent herd of Bretons revolving around themselves, clattering on the spot, tempered by their priests teasing it like sheep dogs". In part fascinated, in part disgusted, he saw the multitude "of cripples and single-armed people, of deformed children […], aged men whose goi-

15 Some letters and accounts of pilgrims are known today that indeed illustrate the attendance of members of the bourgeoisie. In view of the regional structures and measured against their proportion in the population as a whole, the number of rural people and craftsmen might have been the largest, according to current opinion (Schneider 1996a: 275-278; Schneider 1996b: 300-304).

tres are drooping like enormous pears, old women who are limping around […], blind people with pupils coloured white like eggs" and the somnambulistic women who were staring "hypnotised" into the devotional goods within the shops and who were torn away from men whose minds seemed to be "frozen, ruminant like cattle, one does not know why" (Huysmans 1934: 48-51). Huysmans delineated to what extent the stream of pilgrims swamped the streets and the devotional places, making the town "uninhabitable". Around the latrines one could smell "a pestilential taste of human cowpat and urine". Again and again, the crowd was enlarged by new arrivals, "armies of women" approaching "with a din", emitting "raucous cries" (Huysmans 1934: 117-118) and all moving toward the Basilica of *Notre-Dame-de-l'Immaculée-Conception*, itself "a plethora of baseness". Its architecture seemed to be

> "[…] a hemorrhage of bad taste […]. How to define this building whose internal form might remind one, more or less, of the ace of clubs, with five altars each disposed at the top of one of its foliages. One would like to know what style it derives from, for there is some of everything in it, Byzantine and Romanesque, the style of a hippodrome and a casino; but above all, when one looks closely, of a machine depot." (Huysmans 1934: 101–102)

The "masses" of "Les foules de Lourdes" are not the "festive crowd" (*foule de la fête*) joyfully sharing the basic human experience of being part of a community, as described at nearly the same time by Gabriel Tarde (1843-1904) in his essay "Le public et la foule" published in 1898 (Gamper 2007: 480-482). The "masses" of Huysmans are rather akin to the Basilica of Notre-Dame: they represent an amorphous and hybrid entity lacking any balance, harmony, or, so to speak, symmetry. No intrinsic order can be found, no centre and no steady movement. Lourdes cannot be viewed from the standpoint of a central perspective. Strictly speaking, the "masses" do not move. They drift around, directed to the holy places as a multitude of separate units, generally by clerical observers. As a melange of people from different territories, nations, and strata Lourdes also signifies, as Huysmans put it, "the temporary fusion of the castes". Thus, it would be the realisation of the "most audacious chimeras of the philanthropists; at this place the woman of fashion grooms herself, and here the working class as well as the peasant woman cleanses herself; the gentleman and the bourgeois become animals with features of artisans and bumpkins and make themselves to bath attendants while having in mind to be handy" (Huysmans 1934: 196-197).

Especially because of this heterogeneity, the religious "masses" of Lourdes were not a manipulated entity, as the pilgrims of Trier in the writings of Ronge were; they were not an uneducated entity either, waiting for their liberation and emancipation, for it would have been simply impossible for one political force to control their movement thoroughly. They received their energy, if one can call it "energy", from a type of magnetic field creating forces of adhesion and repulsion and suspended between several polarities, such as illness and miracle, vulgarity and the sublime of the sacred, the immediacy of human existence even at its ugliest and carnival-esque masquerade.

It is difficult to say whether the "masses" of "Les foules de Lourdes" could be associated with something that could be qualified as a social counter-point or an analytical counter-concept. They do not fit into a system based only on binary oppositions. One might perhaps think of an army and create relations that might at first glance be called asymmetrical, such as "is more organised than" or "evokes the impression of symmetry much more than". However, in this case, one must also examine contemporary descriptions of armies—and it should not be assumed that the army would have been depicted in terms of order or even of harmony. Huysman's account of the pilgrims of Lourdes is part of an aesthetics of modernity that in the beginning of the nineteenth century began to prefer concepts that could not easily be described in terms of symmetry, such as the "sublime" (Zelle 1995), the "disgusting" (Menninghaus 1999), and later even the "kitchy" (Köster 2006), instead of ideals of balance, concordance, and beauty.

CONCLUSION

To consider societies in terms of symmetry or asymmetry seems to correspond to the tendency to understand human co-existence either as a process of stabilising "order" or as one of creating centres of cohesion by means of a "system". If counter-concepts are considered as a means of constituting and maintaining a divide between "us" and "them" at a given time and within a given society, then they might be evaluated as a tool of establishing order claiming a state of social existence perceived as ideal or simply as normal. In this case, from a logical point of view the respective relationship itself might be called symmetrical or asymmetrical—the problem of how to describe the category of "counter-concepts" more precisely than Koselleck did requires further study. More important, however, is that these types of (asymmetrical) counter-concepts seem to be evoked in particular when protagonists of a supposed ideal state are faced with phenomena they conceive of as simultaneously amorphous and socially meaningful. This is true of the treatment of the "masses" even with regard to historically and systematically heterogeneous discursive contexts. Whether as allegedly apathetic receivers of the mass media or as a social force manipulated by means of religious events, the "masses" very often tend to be described as the "others", and are associated with qualities that create a negative image of a modern and especially bourgeois existence. However, since modernity is hallmarked by a plurality of voices and possible viewpoints, the "masses" are not always addressed when phenomena of great numbers are discussed; the vanishing of the "masses" from the scientific language indicates most commonly a decrease in their political or even ideological dimensions and an increase in analytical precision. Similarly, the "masses" are not always depicted in a way that fits into the binary structure of asymmetrical counter-concepts. This seems to be the point at which the capacity of the "asymmetrical counter-concepts" tool begins to decline and other instruments such as discourse analysis might be more effective.

References

Assmann, J. 2008. *Of God and Gods. Egypt, Israel, and the Rise of Monotheism.* Madison, Wisc. and London: University of Wisconsin Press.

Bär, J. A. 2000. Lexikographie und Begriffsgeschichte. Probleme, Paradigmen, Perspektiven In *Wörterbücher in der Diskussion. Vorträge aus dem Heidelberger Lexikographischen Kolloquium*, edited by H. E. Wiegand. Tübingen: Max Niemeyer, 29-84.

Bartz, C. 2007. *MassenMedium Fernsehen. Die Semantik der Masse in der Medienbeschreibung*. Bielefeld: transcript.

Böckelmann, F. 1975. *Theorie der Massenkommunikation. Das System hergestellter Öffentlichkeit, Wirkungsforschung und gesellschaftliche Kommunikationsverhältnisse*. Frankfurt am Main: Suhrkamp.

Bödeker, H. E. 2002. Reflexionen über Begriffsgeschichte als Methode. In *Begriffsgeschichte, Diskursgeschichte, Metapherngeschichte*, edited by H. E. Bödeker. Göttingen: Wallstein, 73–121.

Bothe, F. (editor). 1920. *Frankfurts wirtschaftlich-soziale Entwicklung vor dem Dreissigjährigen Kriege und der Fettmilchaufstand (1612-1616), p. 2: Statistische Bearbeitungen und urkundliche Belege*. Frankfurt am Main: Joseph Baer & Co.

Brantlinger, P. 1983. *Bread and Circuses. Theories of Mass Culture as Social Decay*. Ithaca and London: Cornell University Press.

Breuer, S. 2005. Ästhetischer Fundamentalismus und Eugenik bei Kurt Hildebrandt. In *Wissenschaftler im George-Kreis*, edited by B. Böschenstein/J. Egyptien/B. Schefold et al. Berlin: Walter de Gruyter, 291-309.

Burke, M. J. 2010. Histories of Concepts and Histories of Ideas. Practices and Prospects. In *Eine Typologie der Formen der Begriffsgeschichte*, edited by R. Pozzo/M. Sgarbi. Hamburg: Meiner, 149-161.

Canetti, E. 1960. *Masse und Macht*. Hildesheim: Claasen.

Dipper, C. 2000. Die "Geschichtlichen Grundbegriffe". Von der Begriffsgeschichte zur Theorie der historischen Zeiten. *Historische Zeitschrift* 270: 281-308.

Dröge, F. 1972. *Wissen ohne Bewußtsein. Materialien zur Medienanalyse*. Franfurt am Main: Athenäum Fischer Taschenbuch Verlag.

Elliott, J. H. 2005. The General Crisis in Retrospect. A Debate without End. In *Early Modern Europe. From Crisis to Stability*, edited by Ph. Benedict/M. P. Gutmann. Newark: University of Delaware Press, 31-51

Engelhardt, W. v. 1949. Symmetrie. *Studium Generale* 2 (4-5): 203-212.

Eßbach, W. 2010. Von der Religionskritik zur Kritik der Politik. Etappen junghegelianischer Theoriediskussion In *Die Junghegelianer. Aufklärung, Literatur, Religionskritik und politisches Denken*, edited by H. Reinalter. Frankfurt am Main: Peter Lang, 41-63.

Föllmer, M./Graf, R. (editors). 2005. Die "Krise" der Weimarer Republik. Zur Kritik eines Deutungsmusters. Frankfurt am Main, New York: Campus.

Frey, D. 1949. Zum Problem der Symmetrie in der bildenden Kunst. *Studium Generale* 2 (4-5): 268-278.

Fuhrmeister, C. 2008. Die Sektion Bildende Kunst der Deutschen Akademie 1925-1945. Ein Desiderat der Fachgeschichte. In *Kunstgeschichte im "Dritten Reich". Theorien, Methoden, Praktiken*, edited by R. Heftrig/O. Peters/B. Schellewald. Berlin: Akademie, 312-334.

Furet, F. 1976. Der revolutionäre Katechismus. In *Die Französische Revolution*, edited by E. Schmitt. Cologne: Kiepenheuer & Witsch, 46-88.

Gamper, M. 2007. *Masse lesen, Masse schreiben. Eine Diskurs- und Imaginationsgeschichte der Menschenmenge, 1765-1930*. Munich: Wilhelm Fink.

Gensbauer-Bendler, U. 1989. Dagobert Frey. Lebensphilosophische Grundlagen seiner Kunsttheorie. *Wiener Jahrbuch für Kunstgeschichte* 42: 53-79.

Grüttner, M. 2004. *Biographisches Lexikon zur nationalsozialistischen Wissenschaftspolitik*. Heidelberg: Synchron.

Grunwald, H./Pfister, M. (editors). 2007. *Krisis! Krisenszenarien, Diagnosen und Diskursstrategien*. München: Wilhelm Fink.

Günzel, S. 2004. Der Begriff der "Masse" in Philosophie und Kulturtheorie (I). *Dialektik* (2): 117-135.

Günzel, S. 2005a. Der Begriff der "Masse" in Philosophie und Kulturtheorie (II). *Dialektik* (1): 124-140.

Günzel, S. 2005b. Der Begriff der "Masse" in Philosophie und Kulturtheorie (III). *Dialektik* (2): 113-130.

Gumbrecht, H. U. 2006. Pyramiden des Geistes. Über den schnellen Aufstieg, die unsichtbaren Dimensionen und das plötzliche Abebben der begriffsgeschichtlichen Bewegung In *Dimensionen und Grenzen der Begriffsgeschichte*, by H. Gumbrecht. Munich: Wilhelm Fink, 7-36.

Habermas, J. 1989. *The Structural Transformation of the Public Sphere. An Inquiry into a Category of Bourgeois Society*. Cambridge, Mass.: MIT Press.

Harris, R. 1999. *Lourdes. Body and Spirit in the Secular Age*. New York: Viking.

Hazard, P. 1961. *La Crise de la conscience européenne. 1680-1715*. Paris: Fayard.

Heftrig, R./Peters, O./Schellewald, B. (editors). 2008. *Kunstgeschichte im "Dritten Reich". Theorien, Methoden, Praktiken*. Berlin: Akademie.

Heil.-Rock-Album. [1844]. *Eine Zusammenstellung der wichtigsten Aktenstücke, Briefe, Adressen, Berichte und Zeitungsartikel über die Ausstellung des heiligen Rockes zu Trier*. Leipzig: Mayer and Wigand.

Heinke, J. 2005. *Die Konstruktion des Fremden in den Romanen von David Malouf*. Würzburg: Königshausen and Neumann.

Hickethier, K. 2004. Ist das Buch überhaupt ein Medium? Das Buch in der Medienwissenschaft. In *Buchwissenschaft – Medienwissenschaft. Ein Symposium*, edited by D. Kerlen. Wiesbaden: Harrassowitz, 39-55.

Hölkeskamp, K.-J. (editor). 2009. *Eine politische Kultur (in) der Krise? Die "letzte Generation" der römischen Republik*. München: Oldenbourg.

Huysmans, J.-K. 1934. *Œuvres compètes, vol. 18. Les foules de Lourdes*. Paris: G. Crés and Cie.

Joas, H./Vogt, P. (editors). 2010. *Begriffene Geschichte. Beiträge zum Werk Reinhart Kosellecks*. Frankfurt am Main: Suhrkamp.

Kaiser, J. H. 1949. *Die politische Klausel der Konkordate*. Berlin: Duncker & Humblot.

Kaufman, S. K. 2005. *Consuming Visions. Mass Culture and the Lourdes Shrine*. Ithaca, N. Y.: Cornell University Press.

Köster, W. 2006. Kitsch. In *Metzler Lexikon Ästhetik. Kunst, Medien, Design und Alltag*, edited by A. Trebeß. Stuttgart and Weimar: J. B. Metzler, 194-195.

Koselleck, R. 1975. Geschichte, Historie. I. Einleitung. V. Die Herausbildung des modernen Geschichtsbegriffs. VI. "Geschichte" als moderner Leitbegriff. In *Geschichtliche Grundbegriffe. Historisches Lexikon zur politisch-sozialen Sprache in Deutschland. Volume 2*, edited by O. Brunner/W. Conze/R. Koselleck. Stuttgart: Klett-Cotta, 593-595, 647-691, 691-715.

Koselleck, R. 1984. Revolution. Rebellion, Aufruhr, Bürgerkrieg. IV. Von der Frühen Neuzeit bis zur Französischen Revolution. V. Die Französische Revolution und ihre zeitgenössische Rezeption in Deutschland. VI. "Revolution" und ihre Gegenbegriffe in geschichtsphilosophischer Perspektive. In *Geschichtliche Grundbegriffe. Historisches Lexikon zur politisch-sozialen Sprache in*

Deutschland. Volume 5, edited by O. Brunner/W. Conze/R. Koselleck. Stuttgart: Klett-Cotta, 689-782.

Koselleck, R. 1988. Abstraktheit und Verzeitlichung in der Revolutionssprache. In *Die Französische Revolution als Bruch des gesellschaftlichen Bewußtseins. Vorlagen und Diskussionen der internationalen Arbeitstagung am Zentrum für interdisziplinäre Forschung der Universität Bielefeld. 28. Mai bis 1. Juni 1985*, edited by R. Koselleck/R. Reichardt. Munich: Oldenbourg, 224-226.

Koselleck, R. 2001. Begriffliche Innovationen der Aufklärungssprache. In *Recht und Sprache in der deutschen Aufklärung*, edited by U. Kronauer/J. Graber. Tübingen: Max Niemeyer, 4-26.

Koselleck, R. 2004a. Historia Magistra Vitae. The Dissolution of the Topos into the Perspective of a Modernized Historical Process [1967] In Id. 2004d, 26-42.

Koselleck, R. 2004b. Historical Criteria of the Modern Concept of Revolution [1969]. In *Futures Past. On the Semantics of Historical Time*, by R. Koselleck. New York, NY: Columbia University Press, 43-57.

Koselleck, R. 2004c. The Historical-Political Semantics of Assymetric Counterconcepts [1975]. In *Futures Past. On the Semantics of Historical Time*, by R. Koselleck. New York, NY: Columbia University Press, 155-191.

Koselleck, R. 2004d. *Futures Past. On the Semantics of Historical Time*. Columbia University Press: New York, NY.

Kott, C. 2008. "Den Schaden in Grenzen halten..." Deutsche Kunsthistoriker und Denkmalpfleger als Kunstverwalter im besetzten Frankreich, 1940-1944 In *Kunstgeschichte im "Dritten Reich". Theorien, Methoden, Praktiken*, edited by R. Heftrig/O. Peters/B. Schellewald. Berlin: Akademie, 362-392.

Kotulla, A. J. 2006. *"Nach Lourdes". Der französische Marienwallfahrtsort und die Katholiken im Deutschen Kaiserreich (1871-1914)*. Munich: Martin Meidenbauer.

Krause, D. 2001. *Luhmann-Lexikon. Eine Einführung in das Gesamtwerk von Niklas Luhmann. 3rd Edition*. Stuttgart: Lucius & Lucius.

Laak, D. van. 2005. *Über alles in der Welt. Deutscher Imperialismus im 19. und 20. Jahrhundert*. München: C. H. Beck.

Lill, R. 1978. Kirche und Revolution. Zu den Anfängen der katholischen Bewegung im Jahrzehnt vor 1848. *Archiv für Sozialgeschichte* 18: 565-575.

Lorenz, K. 2005. Reflections on the Concept of Symmetry. *European Review* 13 (1-2): 3-11.

Lüdemann, S./Hebekus, U. (editors). 2010. *Massenfassungen. Beiträge zur Diskurs- und Mediengeschichte der Menschenmenge*. Munich: Wilhelm Fink.

Marcuse, H. 1968. *Hegels Ontologie und die Theorie der Geschichtlichkeit. 2nd Edition*. Frankfurt am Main: Vittorio Klostermann.

Mazauric, C. Einige neue Wege für eine politische Geschichte der Französischen Revolution. In *Die Französische Revolution*, edited by E. Schmitt. Cologne: Kiepenheuer & Witsch, 89-122.

McManus, I. C. 2005. Symmetry and Asymmetry in Aesthetics and the Arts. *European Review* 13 (1-2): 157-180.

Menningshaus, W. 1999. *Ekel. Theorie und Geschichte einer starken Empfindung*. Frankfurt am Main: Suhrkamp.

Mittelstrass, J. 2005. Introduction. *European Review* 13 (1-2): 1-2.

Müller, J. 1955. *Das Weltbild Friedrich Hebbels*. Halle (Saale): Max Niemeyer.

Negt, O. 1978. Mass Media. Tools of Domination or Instruments of Liberation? Aspects of the Frankfurt School's Communication Analysis. *New German Critique* 14 (Spring): 61-80.

Negt, O./Kluge, A. 1973. *Öffentlichkeit und Erfahrung. Zur Organisationsanalyse von bürgerlicher und proletarischer Öffentlichkeit*. Frankfurt am Main: Suhrkamp.

Osterhammel, J. 1996. Transkulturell vergleichende Geschichtswissenschaft. In *Geschichte im Vergleich. Ansätze und Ergebnisse international vergleichender Geschichtsschreibung*, edited by H.-G. Haupt/J. Kocka. Frankfurt am Main: Campus, 271-313.

Palonen, K. 2004. *Die Entzauberung der Begriffe. Das Umschreiben der politischen Begriffe bei Quentin Skinner und Reinhart Koselleck*. Münster, Hamburg, and London: Lit.

Papenbrock, M. 2008. Anmerkungen zur Geschichte und Methodik der wissenschaftsgeschichtlichen Forschung zur Kunstgeschichte im Nationalsozialismus In *Kunstgeschichte im "Dritten Reich". Theorien, Methoden, Praktiken*, edited by R. Heftrig/O. Peters/B. Schellewald. Berlin: Akademie, 25-38.

Parker, G. 2008. Crisis and Catastrophe. The Global Crisis of the Seventeenth Century Reconsidered. *American Historical Review* 113 (4): 1053-1079.

Rau, S./Schwerhoff, G. 2004. Öffentliche Räume in der Frühen Neuzeit. Überlegungen zu Leitgebriffen und Themen eines Forschungsfeldes. In *Zwischen Gotteshaus und Taverne. Öffentliche Räume in Spätmittelalter und Früher Neuzeit*, edited by S. Rau/G. Schwerhoff. Cologne, Weimar, and Vienna: Böhlau, 11-52.

Richter, M. 1995. *The History of Political and Social Concepts. A Critical Introduction.* New York and Oxford: Oxford University Press.

Rohlfes, J. 2005. *Geschichte und ihre Didaktik. 3rd Edition.* Göttingen: Vandenhoeck & Ruprecht.

Sawilla, J. M. 2004. "Geschichte". Ein Produkt der deutschen Aufklärung? Eine Kritik an Reinhart Kosellecks Begriff des "Kollektivsingulars Geschichte". *Zeitschrift für Historische Forschung* 31 (3): 381-428.

Sawilla, J. M. 2010. Geschichte und Geschichten zwischen Providenz und Machbarkeit. Überlegungen zu Reinhart Kosellecks Semantik historischer Zeiten In *Begriffene Geschichte. Beiträge zum Werk Reinhart Kosellecks*, edited by H. Joas/P. Vogt. Frankfurt am Main: Suhrkamp, 387-422.

Sawilla, J. M. 2011. Zwischen Normabweichung und Revolution. 'Krise' in der Geschichtswissenschaft. In *Krise(n) als Leitbegriff und Erzählmuster in kulturwissenschaftlicher Perspektive*, edited by K. Patzel-Mattern/C. Meyer/G. Schenk. Stuttgart: Franz Steiner, (forthcoming).

Scheitler, I. 2000. Einleitung. In *Joseph Görres, Gesammelte Schriften. Volume 17,4. Die Wallfahrt nach Trier [1845]*, edited by I. Scheitler. Paderborn, Munich, and Vienna: Ferdinand Schöningh, IX–LVII.

Schicha, C. 2003. Kritische Medientheorien. In *Theorien der Medien. Von der Kulturkritik bis zum Konstruktivismus*, edited by S. Weber. Konstanz: UVK Verlagsgesellschaft, 108-131.

Schieder, W. 1974. Kirche und Revolution. Sozialgeschichtliche Aspekte der Trierer Wallfahrt von 1844. *Archiv für Sozialgeschichte* 14: 419-454.

Schiewe, J. 2003. *Öffentlichkeit. Entstehung und Wandel in Deutschland.* Paderborn, Munich, Vienna et al.: Ferdinand Schöningh.

Schlögl, R. 2008. Politik beobachten. Öffentlichkeit und Medien in der Frühen Neuzeit. *Zeitschrift für Historische Forschung* 35 (4): 581-616.

Schmitt, E. 1976. Einführung. In *Die Französische Revolution*, edited by E. Schmitt. Cologne: Kiepenheuer & Witsch, 11-30.

Schmitt, E. (editor). 1976. *Die Französische Revolution*. Cologne: Kiepenheuer & Witsch.

Schneider, B. 1996a. Presse und Wallfahrt. Die publizistische Verarbeitung der Trierer Hl.-Rock-Wallfahrt von 1844. In *Der Heilige Rock zu Trier. Studien zur Geschichte und Verehrung der Tunika Christi. Anläßlich der Heilig-Rock-Wallfahrt 1996. 2nd Edition*, edited by E. Aretz et. al. Trier: Paulinus, 281-324.

Schneider, B. 1996b. Wallfahrt, Ultramontanismus und Politik – Studien zu Vorgeschichte und Verlauf der Trierer Hl.-Rock-Wallfahrt von 1844. In *Der Heilige Rock zu Trier. Studien zur Geschichte und Verehrung der Tunika Christi. Anläßlich der Heilig-Rock-Wallfahrt 1996. 2nd Edition*, edited by E. Aretz et. al. Trier: Paulinus, 237-280.

Schneider, M. 2002. Verfahren der Rezeptions- und Wirkungsanalyse. In *Einführung in die Medienwissenschaft. Konzeptionen, Theorien, Methoden, Anwendungen*, edited by G. Rusch. Wiesbaden: Westdeutscher Verlag, 275-293.

Schönemann, B. 1978. Volk, Nation, Nationalismus, Masse. IX. Vom Wiener Kongreß bis zur Reichsgründung. In *Geschichtliche Grundbegriffe. Historisches Lexikon zur politisch-sozialen Sprache in Deutschland. Volume 7*, edited by O. Brunner/W. Conze/R. Koselleck. Stuttgart: Klett-Cotta, 347-368.

Schröder, S. 2009. *Zwischen Christentum und Islam. Kulturelle Grenzen in den spätmittelalterlichen Pilgerberichten des Felix Fabri*. Berlin: Akademie.

Seibrich, W. 1996. Die Heilig-Rock-Ausstellungen und Heilig-Rock-Wallfahrten von 1512 bis 1765 In *Der Heilige Rock zu Trier. Studien zur Geschichte und Verehrung der Tunika Christi. Anläßlich der Heilig-Rock-Wallfahrt 1996. 2nd Edition*, edited by E. Aretz et. al. Trier: Paulinus, 175-217.

Shimada, S. 2007. *Die Erfindung Japans. Kulturelle Wechselwirkung und nationale Identätskonstruktion*. Frankfurt am Main: Campus.

Sloterdijk, P. 2000. *Die Verachtung der Massen. Versuch über Kulturkämpfe in der modernen Gesellschaft*. Frankfurt am Main: Suhrkamp.

Soboul, A. 1976. Theoretische Probleme der Geschichte der Französischen Revolution In *Die Französische Revolution*, edited by E. Schmitt. Cologne: Kiepenheuer & Witsch, 32-43.

Steinmetz, W. 2008. Vierzig Jahre Begriffsgeschichte. "The State of the Art". In *Sprache – Kognition – Kultur. Sprache zwischen mentaler Struktur und kultureller Prägung*, edited by H. Kämper/L. M. Eichinger. Berlin and New York: Walter de Gruyter, 174-197.

Steinruck, J. 1996. Die Heilig-Rock-Wallfahrt von 1844 und die Entstehung des Deutschkatholizismus. In *Der Heilige Rock zu Trier. Studien zur Geschichte und Verehrung der Tunika Christi. Anläßlich der Heilig-Rock-Wallfahrt 1996. 2nd Edition*, edited by E. Aretz et al. Trier: Paulinus, 307-324.

Stephan, A. 1999. *Emergenz. Von der Unverhersagbarkeit zur Selbstorganisation.* Dresden and Munich: Dresden University Press.

Störtkuhl B. 2007. Paradigmen und Methoden der kunstgeschichtlichen "Ostforschung". Der "Fall" Dagobert Frey. In *Kunsthistoriker und Denkmalpflege des Ostens. Der Beitrag zur Entwicklung des Faches im 19. und 20. Jahrhundert*, edited by G. Eimer/E. Gierlich. Bonn: Kulturstiftung der Deutschen Vertriebenen, 137-153.

Szabó, A. 2000. *Vertreibung, Rückkehr, Wiedergutmachung. Göttinger Hochschullehrer im Schatten des Nationalsozialismus. Mit einer biographischen Dokumentation der entlassenen Hochschullehrer: Universität Göttingen – TH Braunschweig – TH Hannover – Tierärztliche Hochschule Hannover.* Göttingen: Wallstein.

Troll, W. 1949. Das Problem des Schönen. *Studium Generale* 2 (4-5): 259-268.

Wehler, H.-U. 1987a. *Deutsche Gesellschaftsgeschichte. Volume 1. Vom Feudalismus des Alten Reiches bis zur defensiven Modernisierung der Reformära. 1700-1815.* Munich: C. H. Beck.

Wehler, H.-U. 1987b. *Deutsche Gesellschaftsgeschichte. Volume 2. Von der Reformära bis zur industriellen und politischen "Deutschen Doppelrevolution". 1815–1845/49.* Munich: C. H. Beck 1987.

Weyl, H. 1952. *Symmetry.* Princeton, N.J.: Princeton University Press.

Wiedemann, F. 2007. *Rassenmutter und Rebellin. Hexenbilder in Romantik, völkischer Bewegung, Neuheidentum und Feminismus.* Würzburg: Königshausen & Neumann.

Wiehl, R. 2003. Begriffsgeschichte zwischen theoretischem Mangel und theoretischem Überschuss. Philosophische Fußnoten zur historischen Semantik. In *Herausforderungen der Begriffsgeschichte*, edited by C. Dutt. Heidelberg: Universitätsverlag Winter, 81-101.

Wölfl, K. 1960. *Das Heilswirken Gottes durch den Sohn nach Tertullian*. Rome: Libreria editrice dell'Università Gregoriana.

Wolf, K. L. 1949. Symmetrie und Polarität. *Studium Generale* 2 (4-5): 213-224.

Wolf, K. L./Kuhn, D. 1952. *Gestalt und Symmetrie. Eine Systematik der symmetrischen Körper*. Tübingen: Max Niemeyer.

Wurst, I. 2010. Der Keim des Lebens liegt in der Masse. Revolutionen im Staatsorganismus, ca. 1850 In *Massenfassungen. Beiträge zur Diskurs- und Mediengeschichte der Menschenmenge*, edited by S. Lüdemann/U. Hebekus. Munich: Wilhelm Fink, 65-88.

Wyder, M. 2008. Professor Dr. Wolf von Engelhardt: 9. Februar 1910-4. Dezember 2008. *Goethe-Jahrbuch* 125: 364-365.

Zelle, C. 1995. *Die doppelte Ästhetik der Moderne. Revisionen des Schönen von Boileau bis Nietzsche*. Stuttgart and Weimar: J. B. Metzler.

From Asymmetries to Concepts*

KIRILL POSTOUTENKO

"Bei den Hellenen und Barbaren handelt es sich – erstens – um einander ausschließende Begriffe, deren Bezugsgruppen auch im Bereich der Wirklichkeit räumlich trennbar sind."
REINHART KOSELLECK,
VERGANGENE ZUKUNFT (1975)

"Le barbare, c'est d'abord l'homme qui croit à la barbarie."
CLAUDE LEVI-STRAUSS,
RACE ET HISTOIRE (1952)

"The world is richer than it is possible to express in any single language."
ILYA PRIGOGINE,
FROM BEING TO BECOMING (1980)

The significance of asymmetry for all forms of life could hardly be overstated. The most common non-symmetrical relations, egocentrically perceived by human observers as oppositions '*past*'/'*future*' and

* The work on this article was supported at various stages by grants provided by German Academic Exchange Service (DAAD), Kone Foundation (Helsinki, Finland) and the Institute for Advanced Studies (Paris, France).

'*here*'/'*there*', variously contrast life to its alternatives. The moment when the living unity abandons the state of stagnant homeostasis ('*past*') and plunges into uncertain changeability ('*future*'), marks its birth as a dissipative system capable of growth, self-innovation and adaptation to environment. Inversely, the existence of an area at which this system claims the totality (or, more realistically, the supremacy) of operational control ('*here*') proves its inalienable essence as a separate being, whereas all other areas ('*there*') treat this essence—in the best case—as a negotiable appearance.

Semantic egocentricity of prevalent asymmetry markers, obvious in '*here*' vs. '*there*' and somewhat less conspicuous in '*now*' vs. '*before*' and '*after*' (Frazer 1978: 59), obscures the fundamental difference between positions from which asymmetrical observations are made, and the systemic asymmetries themselves.[1] Indeed, imposing human generalizations of experience on processes that do very well without any anthropocentric rhetoric at all, seems to be of questionable explanatory value: whereas the terms '*future*', '*past*', and '*time*' have few direct parallels in physics (Davies 1974: 22), the notions of '*here*', '*there*' and even '*space*' do not rank high in the vocabulary of systems theory. But these impositions, being problematic as a *method*, can be of significant interest as an *object* of research. So far social sciences and humanities devoted considerable attention to the typological examination of differences and similarities between systems endowed with generally high (social world) and generally low (natural world) self-observational capacities (see very useful summaries in Jantsch 1980; Godard/Salles 1991). To be sure, the very distinction on which this dualistic comparison hinges may soon become obsolete: in the genetic code, for instance, the amount (and correspondingly, significance) of self-referential signs turns out to be much larger than has

1 Surprisingly, this difference, initially grasped as a solvable problem (rather than a paradox) in cybernetics (Bateson/Mead 1973; Foerster 1982), has been given the first close look in humanities (see, for example, the summary of narratological findings in Ricœur 1984: 165-188, and a case study in Jauss 1986). Soon after that, social sciences followed the lead and studied the difference between observer and the observed on a micro-systemic level (Maturana 1990, Leydesdorff 2006). Many aspects of these later analyses are highly relevant in the context of the present study and will be addressed in a due course.

been previously thought (see Bauer 2008: 67). However, the main reason for trying to refocus the comparative study of dissipative systems is a shortage of models showing how systemic asymmetries in natural systems set social asymmetries to work. It seems to me that the interrelation between inconspicuous but pervasive *asymmetries*, accompanied by unstoppable pursuit of (and belief in) unattainable *"symmetries"*, may be a helpful conceptual dichotomy in attempting such an adventure.

* * *

Initially it seemed alluring to subsume the standard bipolar classification of asymmetries ("time" vs. "space") under the auspices of this overarching dialectics. However, for the purposes of this inquiry—charting the genealogy of asymmetrical concepts—and because of its rather preliminary nature, such a synthesis would have been unjustifiably complex and abstract. Hence the relations between various asymmetries in nature and society are discussed consecutively—from natural times (and social "times") (I) to natural spaces (and social "space") (II). More specifically, the production of social time by means of creating static images of natural asymmetries (which I call *allopoetic modeling*) is contrasted in the first part to the much more productive (and much less researched) technique of imitating dynamic processes occuring in asymmetrical environment (which I call *autopoetic replication*). Among other things, this change of paradigm means bidding farewell to the traditional notion of *observer* in temporal systems: rather than being a single human being of flesh and blood looking at personal organizer, it turns out to be a pair of two distinct communicative roles (*sender + recipient*) played within social time by relatively strict rules, i.e. referring to specific codes in a specific order at specific time intervals. (These roles are called "Dr. Jekyll" and "Mr. Hide" because they have no identity apart from each other and yet cannot be one).[2]

Schematically, the similarities and differences of irreversibility in various dissipative systems, responsible for the production of "social

2 "I rushed to the mirror. At the sight that met my eyes, my blood was changed into something exquisitely thin and icy. Yes, I had gone to bed Henry Jekyll, I had awakened Edward Hyde." (Stevenson 1886: 68)

time", are summed up in a *Diagram*, with the largest systems on the top and their ever smaller replicas diminishing in the downward direction. Inversely, the *second part* tracks down the production of spatial asymmetries in increasing order: departing from the minimal unit, capable of identity by means of establishing its asymmetrical relation with the environment (a chiral molecule), one proceeds to the ever larger natural and social systems up to humans and their groups. The diversification of communicative codes and practices in human society is seen as one of the main reasons for the ability of *Homo sapiens'* to concurrently rely in social interaction on various spatial asymmetries, which are shown to range from simple orientations ('*body*' vs. '*environment*') to complex evaluations ('*privileged access to own states, beliefs or values*' vs. '*restricted access to other's states, beliefs or values*'). As Reinhart Koselleck's asymmetrical concepts are very likely to be rooted in spatial asymmetries of this kind, it seems pertinent to conclude the second part with the attempt to inscribe his dichotomies into a wider context of social asymmetries.

I

At the very first sight, the so-called "temporal" asymmetry between "past" and "future" looks like the relation between an *asymmetrical system* (for example, "earth", "humanity" or "living being") consecutively changing its states from birth to breakup, and a *symmetrical system* (observer) perpetually referring to its own observational activity (known as "present"). There are at least three common ways of making this asymmetry graspable for the common sense, normally labeled as 'irreversibility', 'stochasticity' and 'causality'. *Irreversibility* is best illustrated *ad inverso* by imagining the stream of "events" within a certain system to be running from the end to the beginning, and testing inversion against previous experience. It turns out that some very ordinary facts of everyday life (such as obtaining warm water by mixing cold and hot one, or merely looking at waves radiating from the spot where a stone went underwater) turn into wildest fairy tales if the order of events is reversed (Horwitch 1987: 59-60; Zeh 1994: 7). (To underscore the absurdity of restoration dreams harbored by the die-hard Russian monarchists, Sergei Eisenstein did actually reel backwards the chronicle documenting the destruction of the monument to Alexander

III in his *October* (1927)).[3] *Stochasticity*, translated into the terms of Aristotelian *Poetics*, may be exemplified in unstoppable progression from "history" ('*what* did *happen*') to "tragedy" ('*what* could *happen*') (*Poetics* 1451b: 30-32). For that reason, the initial stages of a certain process (usually relegated to the '*past*'), are always better reducible to a single version than the later ones ('*future*') at which two and more probable versions compete with one another: social discontent may bring on a revolution or quietly die out, sun light can cause mutation or leave the genetic code untouched, and learning English may result in excellent language skills, average language skills, or no knowledge at all (Bateson 1980: 160). Lastly, *causality* is most commonly understood as the dependence of one event upon another which cannot be reciprocated without significantly redrawing the picture of the universe: for instance, it is definitely plausible that the sun may cause a sunburn, but reverting this causation ('*a sunburn may cause the sun*') borders on insanity (for the more rigorous proofs of causality involving sets of three objects see (Hausmann 1998: 64; Horwitch 1987: 199)).

The proof of asymmetry in dissipative systems appears to be solid, and three asymmetrical facets tend to intertwine and reinforce each other: for example, causal asymmetry presupposes both irreversibility and stochasticity—otherwise "future" could cause the "past", and "real" occurrences would be effects of "probable" causes (Horwitch 1987: 74). Inversely, the stochastic nature of asymmetrical environment depends both on thermodynamic uncertainty and on causal unidirectionality: if general entropy could stall or decrease, and causes were "retrodictable" in the same fashion as their effects are predictable, the world would be completely pre-determined, or, better to say, dead from the start (Watanabe 1966: 547). Nevertheless, it is not immediately clear how asymmetries of open dissipative systems generate "temporal" asymmetries in the social world. The two most common reconstructions of this influence could be called *allopoetic modeling*

3 Eisenstein cunningly highlighted the improbability of restoration (in all senses of the word) by framing it into a "realistic" fiction loaded with symbolism: in the movie, the inversion is mirrored by the "original" documentary hinting at the revolutionary iconoclasm of 1917 in Saint-Petersburg, whereas the actual destruction took place in Moscow four years later (see Goodwin 1993: 84).

and *autopoetic replication*. In the first case, popularized by Henry Bergson and Edmund Husserl, the fluid, continuous, analogously perceived "duration" is converted into a stable, static, and digitally expressed "time" (Bergson 1889; Husserl 1928): for example, the early Egyptians, having shared perceptions of lunar motion and floods, have agreed on a common calendar (Elias 1984: XXXVIII, 22). As this scenario has been extensively studied in various disciplines ranging from philosophy of history to psychology of perception (see, for instance, Arnheim 1992: 35-44), it will be given only a cursory glance. The second practice, on the contrary, remains obscure, largely due to the persistent reluctance of sociologists to see in systemic partitions anything but teleologically charged "social differentiation" (see classical examples in Simmel 1890: 169-170; Mannheim 1922: 41).[4] Small wonder that studies of typological similarities between dissipative processes in nature and society have so far done little to elucidate how asymmetries in natural systems set social temporality to work (see Godard/Salles 1991; Jantsch 1980). Hence autopoetic replication deserves, in my opinion, much more attention than it has received so far, even if its role in producing asymmetrical concepts is rather marginal.

Since *allopoetic modeling* merely simulates temporality, it is unthinkable without suspending the systemic nature of both the observer and the observed. Whereas the former is robbed of self-regulatory properties, and its variously conditioned successiveness is reduced to the crude syntagmatic linearity, the latter conveniently dons the Helm of Darkness, pulling all the strings but pretending to be immaterial, immovable and imperceptible at the moment of observation.[5] As a re-

4 Repackaged, this thesis retains considerable currency in modern social studies: "L'institution est une acte de magie social qui peut créer la différence *ex nihilo* ou bien, c'est la cas le plus fréquent, exploiter en quelque sorte des différences préexistantes, comme les différences biologiques entre les sexes" (Bourdieu 1982: 125). In autopoetic replication, the role of institutions seems to be exactly the opposite—producing similarities "*ex nihilo*".

5 This description glosses over the short-term models (such as tower clock) which tend to repress the analogous perception of *durée*—for example, the fact that wear and tear of the clock mechanism *may* lead to its disintegration—rather than capitalize on it. Without long-term models invariant to sensual perception (and discussed below), such devices are useless: providing illustrations, they fail to assume teleological independence. Indeed,

sult, the real-world *simultaneity* of symmetries and asymmetries within a systemic hierarchy (for example, the fact that a perfect equilibrium on the atomic level would not prevent a swinging pendulum from gradually lowering its amplitude and eventually succumbing to friction (Davies 1974: 76)) is turned into *succession* of ascending equilibrations, embedded in the model by the "invisible" observer: Hegel, for instance, calls "true history" the sequence of the Spirit's self-observations ascending from the individual to the world level (Hegel 1807: 585-586). As for the lifetime of a system, its actual relation to environment is usually turned in allopoetic models upside down: rather than being a *determinant* of the difference between 'before' and 'after', it becomes a *function* of transition from the beginning to the end. This functionality explains the habitual sandwiching of "temporal" narratives between perfect symmetries at the beginning (Garden of Eden) and at the end (Communism) (Löwith 1949: 185).

It is hard to deny that many approximations to the asymmetrical concept of history have been achieved with the help of allopoetic modeling, even if their whole thrust was directed against its stifling framework: Condorcet's "historical table"—a calendar of unstoppable progress stretching into infinity—might be the case in point.[6] Nevertheless, this sort of generalization, eminently useful for the purposes of short-term behavioral coordination, seems to be highly problematic for getting to the core of temporal "asymmetry". The issue with the Paradise is not that it has never existed (who knows?), but that equilibrium

public clocks in 15th century Europe could be mechanical or manual, reliable or not, but these differences had no bearing upon philosophy of time espoused by clock users (see Dohrn-van Rossum 1996: 126).

6 Not only Condorcet's mathematical studies on the "probability of the future" (Condorcet 1783), but his very wording in the Sketch for a Historical Picture of the Progress of the Human Mind reveal probabilistic foundations of his unswerving belief in rationality of historical teleology: "Si l'homme peut prédire, avec une assurance *presque* entière, les phénomènes dont il connaît les lois; si lors même qu'elles lui sont inconnues, il peut, d'après l'expérience du passé, prévoir *avec une grande probabilité* les événements de l'avenir, pourquoi regarderai-t-on comme une entreprise chimérique celle de tracer avec quelque vraisemblance le tableau des destinées futures de l'espèce humaine, d'après les résultats de son histoire?" (Condorcet 1794: 265).

cannot be a starting point for a dissipative system: life could not emerge at the moment when all bodies had the same temperature (Davies 1974: 29), communication would not begin in an environment where everyone had the same information at the same time (Shannon/Weaver 1949: 13; Cherry 1966: 171), and trade would have never developed from the state of equity between supply and demand (Lehmann 1999: 15). In this sense, the birth of society from the first detected (or achieved) inequality of values expressed in manifestly unequal terms ('*ceci est à moi*') remains, for all its grimness, a highly plausible genealogy of human social system (Rousseau 1755: 94). Inversely, the problem with Communism is not that it is cannot be achieved (maybe it can), but that one cannot construct a credible temporal teleology based on the imminence of any state determined a priori, because the beginning and the end of a dissipative system cannot be equiprobable. Apparently, the equiprobability of allopoetic models is rooted in their simultaneity: in principle, the last chapter of any book is as easily available to any reader at any moment as the first one (see Červenka 1994). Under such circumstances, the only chance of dampening the iron determinism of allopoetic modeling is to turn a model into fiction which keeps probability of the whole text at a stably low level: as much as every novel is a history that "did happen", it is first and foremost a tragedy that "could happen". But as representations of temporal asymmetry, the mock calculations of Jules Verne filling the pages of his *Voyage to the Moon* do not seem to fare better than the well-meant calculations of Michel Nostradamus, so usually one resorts to generalized social practices relating symmetrical models to their asymmetrical prototypes.[7] For instance, the confusion aroused by double translation ('*simultaneity of a multilevel system* ['relations between X, Y and Z']' → '*successivity of its narrative modeling* ['novel about relations between X, Y and Z]' → '*simultaneity of the model's material existence*' [book containing a novel about relations between X, Y and Z]) is normally dampened by the artificial highlighting of the middle (asymmetrical) link in the chain, which sequential organization reinvokes the phonetic (presence-based) origins of natural language and thus dis-

7 Statistically, December 2012 is less likely than January 2012, but calendars currently sold in bookstores make no provision for that.

courages non-consecutive perception (Derrida 1967: 17, 31).[8] Taken as a norm, this purely conventional linearity simulates irreversible sequentiality of life: particularly (but not exclusively) in fiction, "poetic faith" (Coleridge 1817: 174) helps to treat yet-unread pages as not-yet-existent and their content probable rather than certain (on the latter aspect, see Iser 1993: 21). As for the upset relation between the dissipative system itself (downgraded to the streak of time in global timelessness) and its environmental surrounding (elevated to the controlling position), it is precisely the unquantifiable time interval between the original and the final equilibriums (represented by the Golden Age, Paradise, Garden of Eden, self-knowledge of the Absolute, Communism and so on) which invokes associations with the steady changeability of *durée* and alludes at stochasticity of history: in Millenarian prophecies filled, despite their pragmatism, with genuine suspense, the major question is not "what?" but "when?" (see Cohn 1970: 13-18).

Still, even the special "rules of use" applied to allopoetic models fail to suppress many of their explicit and hidden symmetries, seriously complicating representation of dissipative systems. Squeezing the complex interplay of certainty and uncertainty in a living system through the bottleneck of a single-code syntagmatic linearity comes at a price, bringing to mind the Goedel-Tarsky dilemma between completeness and non-contradictoriness (Smullyan 1991; Chaitin 2007: 320). Whereas many of those allopoetic models of temporality that claim wholeness as their virtue—for instance, Hegelian and Marxian positive dialectics (see Adorno 1966: 49; Sartre 1960: 115-162; Rosenthal 1998)—are abound with wild generalizations and fuzzy logic, those checked against reality are either self-contradictory or manifestly incomplete: in this respect, Dostoevsky's belief in the reign of Russian god *after* history, or Marx's startling inability to describe the future communist society in any detail come as no surprise (see Postoutenko 2009b: 439; Postoutenko 2010c: 58).[9] Another case in

8 It seems logical that special activities enabling *irreversibility* of communication (such as *breath* and *voice*) happen to have deeper roots within living systems than mechanisms responsible for its *sequentiality* (such as *articulation*) see MacNeilage 1991: 179-180.

9 Since micro-irreversibilities in a living system may only accidentally become opposites (Davies 1974, 20), their "synthesis" appears highly unlike-

point is backward causation: if sacred history (or, on a lesser scale, a classical Hollywood movie) *must* be crowned by a happy end, then it is the last state in the life of the system—rather than the first one—that is random (i.e. causally independent from preceding occurrences), and the certain future may cause the likely past (Löwith 1949: 182). The surreality of this state of affairs was often concealed by retaining some elements of the "natural" causality: as a result, the streams of forward and backward causation were often complementing each other, dead-locking history and producing temporal symmetry, which affinity to fruitless fatalism, persistently dismissed by Christian theologians, was obvious to many earlier thinkers including Aristotle (see White 1981).[10]

No matter how problematic the symmetric properties of the allopoetic models are, the presumption of imperceptible perceiver presents an even bigger obstacle to their representational functioning. Historically, the pursuit of "observation without observer" consisted of downplaying one of the instances causing a paradox: whereas the "absolute observer" culminated in transcendental subjectivity literally encompassing (Johann G. Fichte) or even replacing (Edmund Husserl) the whole world, the "absolute observation", popularized by David Hume and carried by Viennese modernists (such as Hermann Bahr and Ernst Mach) to the extreme, consisted of denying the identity of observing agency altogether (see survey, analysis and criticism in Adorno 1956: 193; Burrow 2000: 147-169; Ryan 1991: 1-11; Postoutenko 2009a: 195-198). But nowhere the plight of the undetectable '*I*' was as

---

ly: even if such direct oppositeness does take place on the same level, it is not obvious how it could go beyond and above mutual cancellation in the manner of Hegelian and Marxian *Aufhebung*.

10 The utopian fiction suffered from over-determination from the late 17th century when the search for an earthly Paradise was abandoned—at least, officially—for good and the only uncertainty left was the interval between the first and the second equilibrium (Delumeau 2000: 154). In Edward Bellamy's novel *Looking Backward. 2000-1887*, the temporal symmetry is reinforced twice: not only is the narrator sent straight to the point of post-historical perfection (112 years after the novel was actually written), but the "end of history", called "a paradise" and disguised as the Garden of Eden, is barely distinguishable from its beginning (Bellamy 1888: 134, 174).

apparent as in the studies of dissipative systems proper: suffice to say that Maxwell's demon flunked his assignment to create equilibrium, or *perpetuum mobile*, within a thermodynamic system because of its own physicality and the resulting entropy growth which his inventor failed to notice (Zeh 1994: 40). The fixation on the adverb "now", brought into mode by Augustine (Johnston 1995) and little by little promoted to the core of allopoetic modeling (see above), appears to be particularly puzzling because its symmetrical self-referentiality in relation to asymmetrical environment is by no means representative, let alone obligatory, for human language in general and its temporal vocabulary in particular. Rather, there is a relative abundance of words which accurately reflect and measure the overlapping between the irreversible processes taking place in the observer and the observed. For instance, '*knowing*' clearly rules out the state of uncertainty in *both* parties involved, because it prohibits not only "observing future", but the "future observer" as well—otherwise "I will know" would have the same informational value as "I know" (Horwitch 1987: 7).

Taken together, the limited representational value of allopoetic models and the apparent abundance of communicative means that reenact (rather than merely illustrate) natural irreversibility in social life, made *autopoetic replication* a rather common mechanism of producing time in society. However, unlike allopoetic models, popular with scholars and ubiquitous on the consumer market (from watches and calendars to astrological fictions), autopoetic replicas have never become a part of common knowledge: for the most part, their studies are tucked in various domains of scholarship from biology and economics to literary scholarship and general systems theory (Maturana/Varela 1980; Neuliep 1991; Hellal/Heddaya/Bhargava 1996: 61-72; Popa 2004: 142-143, 189-192; Lewes 2006). It seems therefore practical to attempt coordination of this scattered knowledge, starting with the most common dissipative systems in nature and then descending to the specifically social temporality.

Fortunately, even a cursory glance at an average bifurcation that eclipses symmetry in a "natural" system suffices for spotting some basic rules of autopoetic replication. Thus it seems evident that a minimally sustainable open system cannot consist of open systems only: in other words, as time goes by, not all parameters of the system but only some—necessarily including the most general, regulative ones—slip into uncertainty (Prigogine 1980: 13). In this respect, the previously discussed changes in gravitational systems which leave molecular symmetries untouched may be attributed a double meaning: the temporal asymmetries located on the level *x* may either override subordinate atemporal symmetries (level *x-1)* or be overridden by the superordinate ones (level *x+1*), but in dissipative systems they will *always* be on the top. This rule appears to govern social systems, too, but its full meaning is not as simple as it sounds. Some (not all) physicists opine that the totality of open and closed systems in the universe does not seem to be irreversible or stochastic (Horwitch 1987: 54-55, 57; Davies 1974: 59), and it might be tempting to crown social world with a similar kind of super-symmetry. But it would be also very inconclusive to yield to this temptation after picking allopoetic modeling apart. That's why (a)temporal symmetry in social systems is treated below as a useful illusion (see Jankélévitch 1974: 114), created by indispensable speed difference between various communicative forms of social self-organization.

Evidently, the survival of any system not only requires a well-developed hierarchy of dissipative and closed subsystems, but also depends on their ability to internally switch between stable and unstable states under environmental pressure. Invariably, such stability, or temporal symmetry, involves extra self-organization (Hollis/Wickman/Korley 1998): when a sudden cold turns water into ice, the resulting crystal lattice displays regularities which *appear* invariant to time and could hardly occur in steadily moving liquids. In a changeable climate, the puddles will freeze and unfreeze many times before summer (or winter) comes, but apparently there is little coordination between various symmetrical states. Such a coordination would invariably require a separate system with a different, preferably a much slower dissipative tempo, so that the sequentially of the changes from one aggregate state to another and back could be observed from a relatively stable standpoint and registered as a single phenomenon. Besides, one would need a measuring device capable of relating the symmetrical macro-states x^1, x^2, and x^3 ('*puddle N was frozen on October 5, 10, and 18*') to the mega-state *X* ('*puddles usually freeze when the temperature falls below 0°*'). Such mega-states, depending on their complexity and flexibility, could be called either *norms*[11] (signs or sign systems which con-

11 As one could notice, the first definition largely follows Parsons' look at *norms* as supplementary mechanisms of behavioral selection allowing for a

tinuously and irreversibly changeable meanings are perceived as unchangeable) or *institutions* (automatized systems, or machines, reproducing inputted norms under environmental pressure or—rarely—as a result of direct external regulation).[12] In its turn, the system operating the measuring device linking micro- to macro-states would be no one else as our good old-fashioned *observer*, stripped from omniscience, anthropomorphic individuality and other remnants of transcendental idealism so persistently cultivated in phenomenologically oriented social sciences.

It would be very natural to inquire how the slower temporality of macro-states is achieved and how the observer that is neither omniscient nor exempt from the second law of thermodynamics can do much better than Maxwell's demon and his modern followers in their attempts to structure asymmetrical systems (see, for example, Davies 1974: 53). The first question could be probably answered by the reference to the non-*indexical* (*symbolic* and *iconic)* codes, which information and even messages are decoupled from the physical presence of communicative agencies and can therefore outlive their messengers or storage devices. Here humans have unquestionable preference over other animals: a lion is able to warn intruders only as long as he is actually clenching his teeth, but a letter full of slanderous accusations may come into force in the absence of its author, and has good chances to outlive both the sender and the recipient(s): an example of the *Protocols of the Elders of Zion* (1903) in Nazi hands is telling enough

far more drastic reduction of social complexity than deliberations (Parsons 1982: 192). But this fairly general view is framed in specific terms compatible with observation of temporality in social systems. As a striking case of a norm which inevitable changeability is stubbornly disregarded for centuries, one could point at a Western notion of "romantic love" which has been still widely considered "eternal", even "natural", despite its late—and purely cultural—origins in the 12th century fiction and constant adaptive changes since then (see de Rougemont 1972).

12 The mechanistic ruthlessness of *institutions* is underscored by the absence (or at least rarity) of feedback trail for norm implementation: the language, for instance, lacks ability to learn whether the words usage in any specific case is correct. In this sense, calling institutions "allopoetic *machines*" (rather than *systems*) could be a forgivable tautology (see Guattari 1993: 87).

(Poliakov 1997).[13] It goes without saying that neither self-organization nor environmental adaptation of a dissipative system require activity of specific communicative units: a person rushing to an exit may just act (i.e. rush) without necessarily sending a message '*I am in a hurry*' (Luhmann 1987: 208), and the long-standing artistic inspiration of a waterfall has no bearing on the dynamics of falling water (Iampolski 1987). But the remarkable longevity of symbolic and iconic codes is rooted not so much in their independence from as in their indifference to individual speakers. In contrast to the mere physical presence of a system, or *indexical* pointing at it (roughly translatable as ['*I am here and now*'] or ['*you are here and now*']), its *symbolic* description ('*Kirill is in Paris on December* 1st*, 2010*') may be sent and perceived from very different angles, and its validity, in theory, is indefinite (Postoutenko 2010a: 272). To be sure, this infinity is as much a hoax as the earlier discussed "stochasticity" of a novel in the course of its reading: in a very distant future, the sentence will likely become imperceptible due to a number of reasons such as the breakup in the chain of semantic modernizations within English language, or the extinction of all storages containing this information, or maybe something else. However, as a dissipative system, natural language is slow enough to be perceived as a "timeless" norm, and even its indexical (deictic) part, as one will see below, is sufficiently impersonal to win the normative status for the whole institution. What remains to be seen is how the observing agency, relieved in the previous paragraph from the necessity of being a living system, manages to link dissipative systems with their normative decelerations.

It has been shown before that an imperceptible observer is an idealist fiction, and a solitary human-like observer could hardly do much more than create additional entropy rather than establish order. The

---

13 In using the terms 'Icon', 'Index' and 'Symbol', I am following the classical triad of Ch. S. Pierce: "An *Icon* is a sign which refers to the Object that it denotes by virtue of characters of his own, and which it possesses, just the same, whether any such object really exists or not. […] A *Symbol* is a sign which refers to the Object that it denotes by virtue of a law, usually an association of general ideas, which operates to cause the Symbol to be interpreted as referring to that Object. […] An *Index* is a sign which refers to the Object that it denotes by virtue or really affected by that object" (Peirce 1903: 102-104).

possible way out of this unfortunate situation is dissociation between the organismic unity of a single living being and its communicative capacity (Leydesdorff 2006): in other words, it takes at least two interaction participants to increase, rather than decrease, its order to bring about organization. An example from physics may illustrate the point: it takes *two or more* waves to create interference effects, and, depending on the timing of waves, the waves may double (in-phase) or cancel (out-of-phase) each other (Cutnell/Johnson 2005: 487-488). Evidently, the second option, for all its harmoniousness, could hardly be called communication (it would be a big stretch to say that two persons singing in unison are "talking" to each other), so the waves should somehow be similar without being identical. In all probability, it is precisely the temporal gap between original dissipative systems and their decelerated replicas (norms) that constitutes this "similar dissimilarity": slow and impersonal enough to be perceived as "stable" (or symmetrical in relation to time), such replicas should at the same time be addressed sequentially in order to generate information and decrease social entropy. Yet this knowledge may generate information rather than noise only insofar as there is a certain process of generating *and* perceiving the utterance "The cat is on the mat" (Luhmann 1987: 198-199): naturally, if the recipient new as much as the sender, he or she would not bother listening or reading. It is precisely the split of a single all-knowing observer into a less than perfect pair of sender and recipient—a sort of Dr. Jekyll and Mr. Hyde—that brings the relation between the smaller and the larger communicative irreversibility in order: the smaller asynchrony ('*speaker*' *before* '*recipient*') ensures the functioning of the larger one (informational imbalance between the '*speaker*' and the '*recipient*' changes *faster* than the normative values of social subsystems).[14] A simple conversation could be a good example of the latter, less trivial asynchrony: for any interaction involving the word "table", knowing that its meaning does not change in the course of conversation is essential (Schütz/Luckmann 1991, 149).[15]

14 Surprisingly, Niklas Luhmann allows for backward causation in communication going from recipient to sender (Luhmann 1987: 198), but this exception seems to me hardly reconcilable with the systemic description of society.

15 Michel Foucault had been one of the first social scientists to recognize that this background knowledge (which he baptized "archive") had an essen-

But other, more complex interaction scenarios such as trials, legitimate power changes etc. leading to creation or upholding of social temporality largely work along this lines: autopoetic replicas (norms) become facts of social life as long as *two or more* living beings, grouped as senders and recipients, perform coordinated consecutive actions which either confirm or negate those norms.[16]

By gradually lowering the observation point *within social dissipative systems, it might be possible to test* the universality of some general rules governing autopoetic replication, as well as to take a closer look at the observer's mediation between social systems, on the one side, and their normative replicas and institutional simulations, on the other side.

* * *

From the purely graphical standpoint, the difference between two representations of dissipative systems—the Prigogine general bifurcation scheme (*Diagram (1)*) and the Illing "tree" of monetary transactions (*Diagram (2)*) is undeniably similar: the fork depicting probable meanings of a regulative parameter is assembled from multiple straight lines: diverging in order to connect the dots denoting transactions, they steadily multiply in the direction of growing stochasticity. However, if the emphasis in the diagram were put on dissimilarities between natural and social systems rather than their affinities, the first fork would have retained the shape of a smooth unbroken hyperbole it had in the original diagram (see Prigogine 1980: 124). Setting the *Diagram (1)* apart from the rest of the table, this difference highlights the vitality of sequential discreteness for communication in social systems: whereas the asymmetrical changes in puddles, pendulums, galaxies and other non-social systems routinely overrule subordinate *low-level* symmetries and are expected to flow more or less continuously and smoothly, each step of the monetary circulation (and any other communicative process) requires consecutively ordered interference of the two-faced

tially systemic—rather than structural—nature : "C'est le système général de la formation et de transformation des énoncés" (Foucault 1969: 171).

16 The fact that norms do not generally rest on compliance but rather on their recognition (including negation) is well-known to sociologists and anthropologists (Popitz 1961: 61; Rappaport 1999: 425).

observer for the purpose of *high-level* pseudo-symmetric (normative) coordination. More specifically, each transaction consists of only one interference (*sender* emits message x at t° → *recipient* decodes message x at $t^{\circ+1}$), but includes multiple references to the autopoetic replicas of natural systems (i.e. norms and institutions). In case of economic exchange (*Diagram (2)*) the motley repertoire may include some very unspecific norms with very slow semantic dynamics (the meaning of the word combination "the U.S. dollar"), some highly specific replicas fluctuating from case to case (a value of a certain currency chosen to be standard, in relation to other goods (Illing 1985: 18)), and many intermediate norms and institutions covering both the last certain ("past") and some probable ("future") steps and aimed at additional equilibration: here the examples range from the exchange of gifts between archaic clans to the various kinds of guarantees in a modern consumer society (Mauss 1924: 151; Bataille 1933; Sahlins 1972: 1-40; Akerlof 1970: 499-500; Spence/Zeckhauser 1971: 380; Haase et. al. 2003: 115; Stöbener 2008: 50-51).

The longer the periods of "suspended asymmetry" are, the more atemporal normative temporality appears to be: particularly meta-norms, which establish, say, the social obligation of exchanging values "fairly" rather than asserting the equilibrating value of each "fair" exchange (or pair thereof) have a strong tendency to look eternal. A case in point could be so-called "turn-taking"— a set of regulations modeled upon the institution of conversation and placed on social settings of different scope involving small and fixed set of designated participants (conversations, adversary trials, democratic political systems etc.). Presented in a *Diagram (3)*, "turn-taking" leaves the normative coordination of adjacent steps intact: only one party at a time talks, or (inter-)acts otherwise (Sacks/Schegloff/Jefferson 1974: 700). At the same time, "turn-taking" relays the selection possibilities in communication from messengers (which remain more or less the same for the duration of many turns) to their messages (which, unlike in "sequential economy", substantially vary within a code chosen by the first user—see Sacks/Schegloff/Jefferson 1974: 701). Effectively, the length of each turn is only restricted by the readiness and ability of sender-recipient pair to keep interaction going and remain a single observer. This combination of restrictive traits (literally coinciding with the minimal conditions for interactive behavior), and breathtaking flexibility (extending to nearly everything else in communication) explains, at least in part, how turn-taking, boasting the longer history than many of its famous institutional equivalents (such as "democracy"—see Schegloff 2006: 71), looks even more stable than it actually is: even in the absence of common language with and societal information about locals, foreigners take for granted that their attempts at communication would provoke a response. In general, turn-taking functions as an ultimate safeguard against communication breakup: if low-level norms (such as topicality of conversation or even friendliness of interactional behavior) fail, the mere participation in a dialogue precludes decomposition of interactional setting (Gavruseva 1995: 343; Ladegaard 2009: 657) or even prevents bloodshed—for example, in the case of robbery (Luckenbill 1979: 100; see also Mauss 1924: 161-162).[17] Compared to the other autopoetic replicas, turn-taking appears to be the closest approximation to the communicative process in general, showing one of its most common abstractions—a pair of communicators—in the flesh. For getting to the core of the intersystem dynamics that creates social temporality, it may be therefore instructive to look at the most basic and arguably original form of turn-taking—a conversation—at a closer range, tracing the intricate relations between the observers' internal and external coordination on the one hand, and the informational value of interactional process on the other.

At the face of it, the restrictiveness of "turn-taking" collides with the notion of dissipative system, but in reality none of the latter's properties is put into question. Indeed, turn-taking is clearly irreversible: *as* in "sequential economy", the informational stand of each conversation's beginning is never repeated. (Even a statement made in a language unknown to the recipient could transmit valuable information, which in some cases—such as the state of war—may decide between life and death ('*It is not one of us*'—see Searle 1969: 44)). It is also undoubtedly stochastic and causal: *unlike* in "sequential economy", the length of each role is increasingly varied, and the senders' elaborations

17 The rhetorical wars with "autocracy" in general and turn-management in particular, waged in science and arts with somewhat more success than in politics, are usually coupled with the declarative upgrading of topical correlation: "Dada ist ein CLUB, der in Berlin gegründet worden ist, in den man eintreten kann, ohne Verbindlichkeiten zu übernehmen. Hier ist jeder Vorsitzender, und *jeder kann sein Wort abgeben, wo es sich um künstlerische Dinge handelt*" (Huelsenbeck 1918: 25).

or observers' responses overwhelmingly depend on the preceding parts of conversations, whereas the conversations' beginnings are usually random, or at least more random than middles and ends.[18] This remarkable flexibility is caused by various properties of symbolic codes ranging from common differentiation between the speaker and its message (discussed above), to the more specific *linearity* (consecutive placement of signs in a message) and *directionality* (synchronization of orders in which messages are generated and perceived by observer): as one goes from higher to lower levels of communicative system, the importance of these properties grows, and they will be invoked below with increasing frequency. However, such seemingly boundless freedom of conversational arrangement would have quickly turned into hopeless cacophony and merge with the surrounding entropy if norms governing turn-taking were qualitatively vague or quantitatively indefinable: unlike a simple coordination between Dr. Jekyll speaking (or writing) and Mr. Hyde listening (or reading), their mutual transformations *(X(sender)/Y(recipient) → Y(sender)/X(recipient) → ...)* are subject to bigger uncertainty and variability known as "double contingency" (Vanderstraeten 2002). In the environment which major characteristics—such as low level of background noise, close (but not too close) interpersonal distance, and face-to-face orientation of potential communication partners—are conducive to interaction, *speaking* of one person would more or less automatically result in *listening* of another, with no special acts of will involved.[19] However, no such automatism produces the changes of communicative role (*speaker* becoming a *listener* or vice versa): the choices between talking back, keeping listening, or going away are all but predetermined, and the other party has surprisingly little control over them. Possibly, it is precisely this unresolvable tension between the qualitative and the quali-

18 It seems like this randomness correlates with the experience of functioning as an observer, amassed by the potential pair of communicators: among friends, it is quite common to engage in interaction without giving or demanding specific reasons for it (Schegloff 2004: 82).

19 The abundance of signs indicating *non*-listening (absence of head-nods, rubbing the eyes, yawning etc.) in the semiotic repertoire of non-verbal communication confirms the suspicion that when Dr. Jekyll speaks in a standard interactional setting, Mr. Hyde's listening is treated as a default (unmarked) behavioral option (see Hargie/Saunders/Dickson 1994: 216).

tative aspects of turn-taking, which makes it an open-ended system and lends it remarkable longevity.

The overarching illusion of (atemporal) symmetry between the sender and the recipient in a communicative act is reinforced in turn-taking at various levels. As in the case of archaic gift exchange discussed before, the time may be believed to "stand still" not only between the sender's and recipient's recourse to norms (which would be in this case semantics and syntax of natural language) but also between the replicas of two observers swapping their roles. The latter is particularly noticeable in interrogative constructions ('*How are you?*') followed by answers ('*Not bad*'), which otherwise would be of no epistemological value (Speas/Tenny 2003: 320-321).[20] Nevertheless, these and other forms of symmetrical parity, not very strong from the start (the response time in a dialogue is usually limited), are further undermined by the speaker's right to select himself (or herself) for the next turn (Sacks/Schegloff/Jefferson 1974: 701): indeed, the sender can in principle go on forever, and the recipient has few resources to cut him (or her) short before the external limitations bring interactional setting to a close: in this way, Cato the Younger prevented voting on measures he opposed by talking in Roman Senate until dusk (when the sessions had to be ended by law (Loewenstein 1973: 210)).

This old example (and its modern repercussions, such as the so-called "filibuster" in the US Senate) show, however, that the unchecked freedom of self-selection in turn-taking may eventually lead to the disintegration of observer and the breakup of communication chain. Preventing response at all cost, the absolute self-selection comes at a price. Firstly, the content undergoes massive devaluation:

---

20 Another, more general example of communicative practice dependent on imagined (a)temporal symmetry is the so called "repair" (Goffman 1971: 95-187; Schegloff/Jefferson/Sacks 1977): usually illustrated by the apology following deviant behavior, it is as clearly discernible in turn-taking on a larger scale from confessions to indulgences (Hahn 1986). But the goal of restoring status quo may only seem realistic if there is possibility of applying the repairing procedure directly to the damaged state, which in manifestly impossible, as the reaction time of the "offender" cannot be 0. That's why some non-democratic systems, having—quite literally—little patience with their offenders, reject this norm outright (see Postoutenko 2010b: 32).

under given circumstances, it has no meaning except for grotesquely exaggerated linearity and directionality of natural language. Secondly, the recipient, whose attention is demonstrably ignored and whose "turn" is unceremoniously relegated to the hypothetical future only marginally related to the current exchange, is nearly pushed out of the communicative process. Such practices, resembling the trickery of some pre-Julian emperors whoprolonged their times in office by stretching the respective calendar years (see Heitland 1909: 346), also shedlight on communicative deficiencies of authoritarian systems with their clueless leaders, obsessed with self-performance, majorities at any cost and endlessly multiplying feedback chains (Rosenfeldt 1991: 145; Postoutenko 2010b: 28). The only normative practice preventing such abuses is pre-allocation of terms, common for institutional settings (Sacks/Schegloff/Jefferson 1974: 729), but its overzealous application—as the morbid stability of some rituals seems to attest—may only make things worse: a ready-made script of a dialogue would effectively reduce the living openness of autopoetic replica to the deadly fictionality of an allopoetic model. No real observer is possible if the relations between the "fast" (interaction) and the "slow" (norms) dissipative systems are regulated by the sequential orders unrelated to the semantic development of either regulator: if listeners are ordered to give a standing ovation to Joseph Stalin and to shout "Long live our great leader" at the very end of his speech *regardless of its content*, the coordination between the sender and the recipient—if any—would actually be a noisy nuisance, an unwelcome interference with the ceremonial procession, whichharmony is predetermined by backward causation of its script (Postoutenko 2010b: 24-25). Given those problems, the turn-taking in modern societies presents at the first sight a somewhat relaxed version of sequentiality observed above in the system of monetary transactions. On the one hand, a more or less loose combination of norms ('*X and Y should have roughly similar times in conversation*', or 't^n *is a suggested length for a turn in conversation [in elected office]*') encourages (but rarely enforces) the smooth, rhythmical flow of communication in order to create in the eye of the beholder symmetries which may be simple (*sender [power] ↔ recipient [opposition]*) or complex (*sender [power] turned recipient [opposition] ↔ recipient [opposition] turned sender [power]*) . On the other hand, this freedom, indispensable for the systemic life of social systems and their subsystems such as language and politics, is occasionally restrained by the externally imposed rules dictating pre-allocation of turns in institutional settings (‘*The speaker will give a talk for about half an hour, and then every participant will have 15 minutes to respond*’): here the allopoetic models of public performance are disguised as communicative events but obviously formulated in fictional terms extraneous to communicative performance. In all probability, this unpopular and moderately efficient but tolerated “external regulation” (Elias 1976: 48) emerged as an imperfect response to the incomplete self-organization in a system which open-ended sequentiality extends beyond the level of transaction to the semantics of each message.

Last but not least, it is worth looking at a micro-level of social temporality, at which the observer’s integrity is taken for granted and the only sources of irreversibility, stochasticity and causality are linearity and directionality of a preferred symbolic code (natural language). Traditionally, this level is broken down into larger (narratives) and smaller (sentences) units regulated respectively by plots and grammatical rules.[21] In narratology, the transition from allopoetic models to autopoetic replicas, being still in progress, is particularly visible in juxtaposition of the early structuralist morphology of folktales (Propp 1928: 129-148) and its probabilistic revision half a century later (Bremond 1973: 131-134). In the place of backward causation, typical for fictional *texts* (see above) and promoted by Vladimir Propp to the regulative principle of *narrative system* (which renders his own term “sequence” senseless), one finds in Claude Bremond's “logic of narration” the familiar tree of radiating possibilities, determined, unsurprisingly, by textual references to unstable initial conditions (“eventuality”) and protagonist’s (trans-)actions as “positive” (‘*transition to action*’) or negative (‘*non-achievement of a goal*’) consequences of these instabilities (*Diagram (4)*).[22] As for the norms regulating narration as a dissipative system, they remain in the shade of the ubiquitous allopoetic model—the symmetrical finiteness of literary, legal and other texts expected to be the end-result of unstable oral or written performance in discourse- and print-oriented cultures. Small wonder that one of the first apologies of *textual* (as opposed to the

---

21 The difference, to be sure, is contested from both sides (see, in particular, the structuralist conjectures: Todorov 1969; Jakobson 1960).

22 This tectonic change in narratology has been studied in depth by Paul Ricœur (Ricœur 1984: 78-87).

narrative) semantics coincided with the explosive growth of book circulation in Europe (Luther 1530).[23] On the other hand, the interest in emancipation of 20th century iconic codes from their indexical content has retroactively provoked interest in paradoxical stabilization of narratives (Benjamin 1935: 477): while their symmetrical externalizations were cast for centuries in stone, parchment, paper and other materials which high durability emanated the feel of temporal symmetry, the strategy of the last centuries has been the shift from the most conservative physical systems (such as stone) to the complex institutions with the best regeneration potential (such as print) (see Innis 1950, 216; Giesecke 2006).

In grammar, on the contrary, the immediate observability of analyzable sequences makes the reflection upon their normative regulation much easier. Unlike stories, which linearity and directionality could hardly be described—as Bremond's model shows—in terms independent from sequentially organized social action serving as a denotative background for narrative temporality (Scholes/Kelogg 1966: 82, 207), sentences and their groups enact the dissipative nature of language through specific grammatical means uniformly traceable—if not immediately viewable—in all sorts of discursive performance. However, this transparency is complicated by the unusually high degree of hierarchical differentiation within the system. While looking at the irreversibility of human speech, which stems, as has been discussed before, from its original symbiotic unity with action, preserved in oral form and imitated in writing, this complexity is not really striking. However, the breakdown of this unity, superordinate to communication, into discrete sequentiality of speech, simultaneously translated in written language into spatial allegories of linearity and directionality (Di Sciullo 2003: 1), reveals a stunning array of consecutive orders refusing to melt together. The arguably best illustration of this state of affairs is provided by the uneven stochasticity of human language. Thus the sequences of letters in a word, the sequences of words in a sentence and the sequences of sentences in a text produce different trees of possibilities oscillating between near-stability at the end of

23 The Romantic cult of the fragment (Frank 1984), half-heartedly revived in Modernism, may be seen as a backlash against proliferation of symmetrical structures in European Classicist fiction (on the latter, see Greenberg 2009: 10).

each low-level sequence ('*L-o-n-d-o*' will be almost certainly followed by '*n*') and rapidly soaring uncertainties in-between at high levels (the exact sentence that will follow '*The cat is on the mat*' in the middle of a story is very difficult to guess). But even the irreversibility of verbal language is constrained by specific norms such as syntax allowing for non-linear subordination in a sentence. It seems apparent that the sentence tree '*This idea is…*', presented in full in a *Diagram (4)*, is almost unconditionally irreversible on the level of verbal semantics; the information value of a sentence gradually increases from subject ('*idea*') to its first (respectively '*interesting*' or '*boring*') and second (respectively '*wrong*', '*productive*', '*unnecessary*' or '*harmless*') predicate.[24] Nevertheless, the very same sentence exhibits partial reversibility, or symmetry, on the level of syntax: minor differences aside, the statement '*this idea is boring and unnecessary*' is similar to the statement '*this idea is unnecessary and boring*'. Clashing with the narrative linearity, the same kind of forking comes to light in anaphoric chains, roping sentences together in much the same way as our sight mends enfilades: thus, in a three-sentence group '*The author of this article is X. He is also Y, and he is Z, too*' the pronouns '*he*', directly subordinated to the proper name at the beginning of the text, may be easily substituted for each other (Marácz/Muysken 1989: 29). Both forks look like classical examples of causality: predicates cannot govern subjects, and neither first nor second '*he*' can exchange places with '*the author*'. At least as far as syntax is concerned, causality emerges as a stronger protector of systemic openness than irreversibility, despite its indulgence to secondary symmetries such as syntactic parallelisms.

Unsurprisingly, the degree of self-differentiation within natural language weighs on the diversity of autopoetic replication produced by

24 Of course, verbal language is fully capable of producing perfectly reversible signs: the medieval saying "*in girum imus nocte et consumimur igni*", (mis)attributed to devil, reads both ways without a slightest difference. Nevertheless, it seems evident that no random information will survive confinement in such a procrustean bed. With some sporadic exceptions such as advertisement (Pfeiffer 1985), the textual symmetry, rather than being a means of reducing narrative complexity, turns in palindrome into *the* end-in-itself (see Rusek 2009). As a result, its meaning becomes utterly irrelevant, being in most cases just barely above the threshold separating information from noise.

its various subsystems. Coupled with the capacity of the linguistic system to act at the forefront of social life as *the* institution (re)generating norms, the self-organization of human speech reduces complexity of some norms. For example, the meanings of single words and utterances, which popped up in previously discussed contexts as scattered accidentals ('*the U.S. dollar*'), turn out to be much less arbitrary if put on the crossroads of various paradigmatic relations accessible to average language users such as '*the U.S. dollar*'—'*the Australian dollar*'—'*the Canadian dollar*', or '*one dollar*'—'*two dollars*'—'*three dollars*', etc. Other autopoetic replicas—for example, the frequent reversibility of propositional statements ('*Kirill is the author of this text*' = '*the author of this text is Kirill*') are smuggled from symbolic codes with negligible social referentiality such as logic or mathematics, where they serve as generalized allopoetic models of abstract relations: as the unchangeable mental "denotates" of symmetrical statements such as '*X = Y*' are projected upon living systems ('*Kirill*') and their social replicas ('*the author of this text*'), the illusion of non-changeability suppresses awareness of the latter's dissipative nature. But arguably the most interesting norms come forth in the process of translating high-level pseudo-symmetries into specifically linguistic terms. For example, in Totonac-Tepehua languages the penchant for social equality, or symmetry, renders senseless the difference between two diametrically opposed versions of an ordinary transaction between sender and recipient: the crucial difference between '*I promised you money*' and '*You promised me money*' is nullified in their conflation '*We promised each other money*', which diplomatically merges two radically different constructions of observer into a single first-person plural pronoun (Beck 2001: 60-61; compare Fairhurst 2007: 109-110).[25]

One may notice that the resulting sentence, impeccably egalitarian as it is, makes little sense as it ignores the temporal character of communication at two levels at once. Firstly, the mere contact between the sender and the recipient (or, in other terms, any transaction between creditor and debtor) requires their consecutive coordination, which

---

25 It is interesting to compare this syntactic symmetry with the symmetrical lexicon of at least some Papuan languages (such as Toaripi or Namau) which used the same expressions for referring to buying and selling (Holmes 1924: 294-295). Unsurprisingly, Marcel Mauss found these facts very relevant for his exchange theory (Mauss 1924: 193).

rules out any temporal symmetry. So when the sender promises money to the recipient, the latter *must* read (or listen) rather than speak (or write), and then accept (or reject) this promise rather than promise himself (Luhmann 1987: 203). Secondly, the verbal contract between the creditor and the debtor cannot take place—as has been ascertained before—under conditions of equilibrium distribution of value, as its purpose is the advance guarantee of such equilibrium covering the period between *this* and *the next* transaction of the same communicators (the actual disbursement of money). In contrast, the mutual promises of one and the same resource, expressed by the sentence conflation '*We promised each other money*', indicate its equal supply on both sides. These peculiar starting conditions not only reinstate the phantom of "original equilibrium" but also describe the whole transaction as a "change" from one symmetry to the other, similar one, which is hardly reconcilable with the notion of dissipative system in general and with the idea of communication in particular. In principle, it is not inconceivable that X and Y may promise money each other, but they cannot do it in a time-symmetrical fashion simulated by the polite Totonac-Tepehua speakers. At the very least, both the promises (as messages) and the guarantees (as social facts) should be sequentially ordered: an irreversible change should take place between the *real* times $t^{\circ u}$ (when I utter my promise), $t^{\circ u+w}$(when you perceive it), $t^{\circ(u+v)+w}$ (when you utter your promise) and $t^{\circ(u+v+w)+x}$(when I perceive it), and a similar lag is expected between *possible* times $t^{\circ\, u+\ w+y}$ (when I consider it probable to give you money) and $t^{\circ(u+v+y)+w+x+z}$ (when you consider it probable to pay it back). Moreover, if the norm establishing symmetry between the steps (bonds of friendship, tradition of gift exchange, dogmatic prohibition of interest rate in Islamic and early Christian societies etc.) is missing, there is a very good chance that the amount promised at $t^{\circ v}$ and at $t^{\circ(v+w)+x}$ would differ, as the higher-level norms that make guarantees comparable in the first place, ascribe atemporality to their *functions* ('*money*' = '*standard of value at* $t^{\circ v}$' = '*standard of value at* $t^{\circ\, v+x+y+z}$') and let their specific *meanings* float free, reflecting in their dissipative micro-dynamics the macro-dynamics of economic system.

* * *

The main goal of this small-scale analysis was to provide a concluding illustration of how various kinds of dissipative changes in social systems find their social expression in regular dichotomies of *temporal (a)symmetries* (representing "fast" communicative process) and *(a)temporal symmetries* (representing "slow" norms and institutions). However, the hidden purpose behind this deconstruction was also to challenge incontestability of *temporal* interpretations pertaining to asymmetrical processes in the social world, and suggest, so to speak, a second, *spatial* axis around which social asymmetries may revolve. In fact, the improbability of (informational) equilibrium as a starting point for a dissipative system, demonstrated here *ex post* by unfolding communicative process backwards, has a much better and simpler explanation unrelated to consecutiveness and sequentiality of any kind: unless information (or, in a more general sense, energy) differs in different parts of the system, no systemic dynamics could be expected. In principle, that gives us the second answer to the question "Why at a certain point Dr. Jekyll knows something that Mr. Hyde does?": if social interaction is to take place, a message must become a social act, which presupposes not only a temporal gap between Dr. Jekyll sending a message and Mr. Hyde deciphering it, but, first and foremost, a cognitive gap between Dr. Jekyll being able to produce certain information and Mr. Hyde unable or unwilling to procure it on his own. Evidently, this unequal distribution of knowledge cannot be fully attributed to Dr. Jekyll's being first to the finish every time, since in this case he would be the only source of information for the whole society, the only speaker around and so on, which even for medieval societies dominated by "silent majorities" would be an unworkable communicative arrangement (see Gurevitch 1990). Rather, one could speculate that Mr. Hyde, too, has some potential to generate information on which his counterpart would never get his hands. In other words, the informational disequilibrium constituting society by means of communication is likely to contain irreversibilities which are neither consecutive *per se* nor can be sequentially ordered, although it is not unreasonable to suppose that they are reconcilable by means of certain pseudo-symmetrical norms (produced by autopoetic replication) in more or less the same way as temporal asymmetries. In my view, it is this sort of cognitive irreversibility, present in most of the animals, that plays major role in producing spatial asymmetries in the society. Furthermore, I would argue that this "spatial" kind of asymmetry may be more important for creating and preserving the unity and identity of living and social systems than its "temporal" counterpart. The fact that spatial asymmetries have a decisive impact on producing asymmetrical concepts in human language might serve as a confirmation of this superiority, which justifies its discussion in this context no less than the topic of the present volume.[26] Following familiar pattern, the next chapter will focus on the genesis of spatial asymmetries, their functioning on the crossroads of large- and small-scale dissipative systems mediated—expectedly—by the observer, and their presence at all levels of social life from interaction in general to a single utterance.

II

The special theory of relativity holds that our universe, unlike its everyday simplification, does not display fundamental differences between its "time" and "space" (see Einstein 1920: 60-61). Consequently, there is little reason to look for an unbridgeable gap between "temporal" and "spatial" asymmetries in social communication: rather, it seems, a much better return could be reaped from investigating those relations in natural and social systems which permit concurrent interpretations of their irreversible properties. The deductive approach to such relations is usually focused on such universal phenomena as gravitation: being on a macro-level the moving force behind such dissipative systems which irreversibility is unquestionable (see, for example, the all-devouring "black holes" (Zeh 1994: 1-2)), it lays on the small

26 Some of Reinhart Koselleck's remarks make an impression of his subscribing to this position: "Hinter der akuten Zuspitzung steht freilich ein allgemeines Oppositionspaar, nämlich den Gegensatz von Innen und Außen, der die geschichtliche Räumlichkeit konstruiert. [...] Und dieselbe Ihnen- und Außenposition taucht in allen Geschichten auf, auch wenn die Handlungseinheiten höher aggregiert sind als nur durch zwei Personen" (Koselleck 2003: 104; see also Koselleck 1994: 276). In similar but somewhat diffuser terms, Niklas Luhmann treats "the difference between Hellenes and Barbarians" as "the difference between exclusion and inclusion" (Luhmann 1995:138).

scale the foundations for top-bottom asymmetry by determining the shape of smaller objects (living beings) in relation to a larger one (earth) (see Morton 1926). It seems very probable that this supplementary contradistinction between the stably lower and the stably upper parts of the body has eventually lead to the progressive functional differentiation of its parts. For instance, the extremities were first diversified on the vertical level, with the *lower* paws ascertaining erect posture and the *upper* paws interacting with the environment, and then on a horizontal level, with the *right* "hand" grabbing foodstuffs and the *left* climbing trees (MacNeilage 1991: 178-179). It is the latter difference which is often regarded as a source of deontic asymmetries in the social world: as long as the majority of human societies associates "right" with "good" and "left" with "bad", the hand that is well-versed in complex operations is held to be a synecdoche of the systemic Self—residue of "goodness" and "rightness" opposed to the essential "wrongness" and "badness" of the environmental Other (which is represented, in turn, by the other, less useful, hand—see McManus 2002: 361-362; Ivanov 1983: 37). Small wonder that under such circumstances the opposition, say, '*Hellenes*'/'*Barbarians*' is often seen as yet another reflection of the closed symmetrical bipolarity that breaks natural and social world into 'men' and 'women', 'yin' and 'yang', 'angels' and 'demons' and other constitutive dichotomies living in every pore of society and its communication (Ivanov 1974: 38). In all probability, this structuralist vision is responsible for seeing asymmetrical concepts as pairs (Koselleck 1975: 212)—a sweeping statement for which, to my knowledge, little argumentation has been provided so far, except for the vague references to the *harmonie universelle* (see Luhmann 1995: 138-139).

In any case, there are many reasons to treat the "right-hand" explanation of spatial asymmetry with caution. First of all, the startling incongruity between the natural phenomena observed on one side, and their social echoes on the other, arouses suspicion bordering on disbelief. It has been shown before that social "temporality" was created by a twofold replication of irreversibility, stochasticity and causality of "natural" dissipative systems, glued together by the observer's coordinated interference. Hence it is natural to expect the process of comparable complexity to be at the roots of social "space". Instead, one is lead to believe that a social subsystem with strong *asymmetrical* features ('*Hellenes can call Barbarians such, but Barbarians cannot*';

'Hellenes are usually "bad" for Barbarians but "good" for themselves' etc.) is born from the *symmetrical* binary model which is demonstrably reversible (left-handers are abound) and plays no regulative role in a system, being clearly superseded by the bodily symmetry: it seems obvious that the sheer presence of both hands is infinitely more important for the systemic survival of humans than the difference between them (Neville 1976: 6-7)). Admittedly, the social universe contains non-temporal binaries with clear differentiation between the marked and the unmarked poles (there is no way of characterizing Christian or Platonic Evil without marking the Good first (Auget 1977: 33-34), but the examples preferred by structuralists—such as *yin* and *yang*—display almost limitless reversibility (Black 1989: 70). All in all, the argumentation for deriving spatial asymmetry of social world in general and asymmetrical concepts in particular from the functional asymmetries of its inhabitants' bodies seems to rest on a fairly uncertain ground. Therefore returning to the spatio-temporal continuum at the bottom of systemic hierarchy and the subsequent inductive ascension to the complexities of the systemic appears to be a more practical option, which, by the way, gives chances for a far more promising interpretation of handedness than the one presented above.

Looking once again at the molecular construction of living organisms, one could fairly easily spot the level at which non-temporal asymmetries start playing regulative roles. Threatened by breakup and dissolution in the hostile environment, living systems are prone to mounting sophisticated defense, concurrently mobilizing their symmetrical and asymmetrical layers in the fight against chaotic surroundings. Roughly speaking, those physical threats that do not subvert clear spatial distinctions and deform systems through outside pressure (rather than from within) are often handled by symmetries: although living systems, unlike water, are unable to turn into perfect cubes (ice crystals) or spheres (drops of liquid), most of them are endowed with fairly stable symmetries of external form, which usually helps surviving winds, gravitation and minor collisions with other bodies (Neville 1976: 6-7). Conversely, the chemical dangers, aimed at infiltrating systems in order to break their self-regulatory chains, are best put off by asymmetries inherent in (if not inseparable from) systemic identities (see details: Insel/Roth/Price 2002: 294). Indeed, in the systemic organization of living organisms, observed from molecular level upwards, the compulsory (self-) differentiation from environment is inextricably tied to asymmetry. Thus chiral molecules, which irreconcilably asymmetrical shapes are similar to the variance between a left and a right hand—but *not* between their functions (!),—are bundled in non-racemic mixtures containing unequal numbers of "left" and "right" enantiomers (Neville 1976: 3). Such a thoroughly asymmetrical structure not only allows for identification of living Self with a non-reversible order of simultaneously present elements, but also permits extending this irreversibility to the topology of outer space, producing a "consistently biased response" to environmental pressure (Bradshaw 1993: 33): in famous experiments of Louis Pasteur, a fungus could tell L-tartaric acid, common for living organisms, from its *de facto* mirror reflection, D-tartaric acid, usually found in dead corpses (Pasteur 1849). The same procedure of identifying Other as non-Self and keeping it off is performed by immune system of an individual (Jacob 1982: 54), although the sheer complexity of the task on the level of individual being may be at times overwhelming (see Anspach/Varela 1994; Pradeau 2009: 337). Crucially, the immune system, being a subdivision of systemically integrated physical body, incorporates to a large extent its internal hierarchy of values based, very roughly speaking, on concentric sequential asymmetry reminiscent of the ancient city-fortresses. The first squads of invaders that have just managed to sneak through physical borders are promptly handled by "innate immunity", which identifies aliens on the basis of non-reflective (inborn) differentiation between "self" and "non-self": the Barbarians, at this stage, are those who '*come from elsewhere*', or '*have always been such*'. However, the more daunting job of unmasking "non-self"—the antigens which have transgressed body boundaries in large numbers and made themselves comfortable deep inside—is done by "adaptive immunity" (Pinchuk 2001: 182-198), which ability for semantic differentiation is mostly based on negative self-affirmation (see Ishida 1996). At this stage, the Other, "bad" and "wrong", is identified as something which does not partake in the consistently biased response going from the center of the immune system (Stewart/Coutinho 2004): Barbarians are Barbarians because they are '*not Us*', or '*not the one who is charged with telling Self from Non-Self*'. It is likely that the system of deictic (asymmetrical in relation to speaker) spatial terms such as '*I*', '*close*' and '*here*', as opposed to '*(s)he*', '*far*' and '*there*', merely translates these relations between human system and its environment (other systems included) into communicative terms, and the basal '*I*' vs. '*(s)he*' contradistinction is often considered to be the main source of epistemic ('*true*'/'*false*') and deontic ('*right*'/'*wrong*') demarcations (Tugendhat 2003: 44; Chilton 2004: 59).[27] For me, the latter view is a plausible alternative to the structuralist genealogy of spatial symmetry, and the only missing link in the hypothesis is the specific communicative coding of systemic identity that leads to formation of "spatial" (as opposed to "temporal") asymmetries in social world.

In my view, the source of the gaping divide between "here" and "there" lies in the abandonment of phased sequentiality in relations between the human system and its environment, common, as one could see, for the immune subsystems of living beings. This break with graduality occurs at the reflective level of systemic self-identification, which is responsible for grasping the system as a single whole. Understandably, for such purposes the multilayer identity of a living being has to be crammed into the stern dualism of impenetrable self-reference ('system-for-itself') and its social digest ('system-for-others'): "internal" systemic states (beliefs, wishes, fears and intentions), genuine but generally incommunicable, are opposed to their "external" environmental projections (cries, performances, claims, warnings, statements), communicable but generally falsifiable (James 1905: 61; Tugendhat 1979: 33; Davidson 2001: 38; Elias 1970: 128-129; Fodor 1975: 82).[28] In reality, the propositional content of someone's belief may well be mistaken, but the very believing may not; conversely, the cries for help—as the Aesop's famous fable *The Boy Who Cried Wolf* attests—may be perceived as lies even if they are genuine. In other words, the validity of virtually all utterances of a certain individual may well be contested by the community of his or her interaction partners, but the privileged access to own states constitutes a precious little piece of everyone's absolute knowledge that others could not judge or partake in: in Wittgenstein's flashy formulation, the meaning of "I feel pain" could be better communicated by action

---

27 I sidestep here the complex question of interrelations between temporal and spatial asymmetries in the semantic structure of the first-person personal pronoun.

28 Remarkably, both Donald Davidson and Ernst Tugendhat call this opposition "asymmetry".

(piercing the recipient with the same nail) than by language (Wittgenstein 1951: 24).

Obviously, such a distinction, however categorical, may not be enacted and sustained without multiple recursive moves between personal thinking and social life. In many (if not most) cases, the awareness of own condition is framed in terms of "external" vocabulary ('*illness*' being a textbook example—see Conrad/Barker 2010); in reverse, many universals—for example, the pronoun '*I*' or the noun '*love*'—would have lost much of their sense if the non-sharable components of their meanings ceased to exist (Molho 1997: 10). Hence the familiar picture of reconciliation between the living system (with its hierarchies of symmetries and asymmetries) and its social replicas (with their asynchronous "slow" and "fast" asymmetries) gets transposed from sequential (y *after* x) to complementary (y *next to* x) terms. This necessarily entails the fundamental change in practices of autopoetic replication. Whereas the temporal asymmetry is based on suspension of *consecutive irreversibility* in "*slow*" norms and institutions (language in a certain country) as opposed to the "*fast*" interaction, its spatial correlate is built upon ignoring *concurrent irreversibility*, or incompatibility, of users' positions in relation to "*global*" norms and institutions ("time" in a modern world)[29], compared to the uniqueness of each such position in every "*local*" pair of users.[30] There is little doubt that the meanings of words '*I*' and '*you*', or '*here*' and '*there*' must be both stable and instantly changeable: as markers of a certain qualitative difference between senders and recipients, they always mean the same, but as references to the specific persons in interaction, they rotate after every turn (see Tugendhat 2003: 44; Postoutenko 2010b: 27). Accordingly, the observer of "spatial asymmetry" is split into two asymmet-

29 Niklas Luhmann goes as far as to link globalization of time after 1850 to the decrease of conceptual asymmetries (Luhmann 1995: 141).

30 To be sure, the positional invariance of symbolic signs is no more than *a* prerequisite for their generality (let alone such phantoms as "universality"), but this difference is eagerly ignored: the majority of language users display considerable benevolence to such manifestly fraudulent statements as "*Everyone knows that X*", and many sociological references to "common knowledge" such as *Wissensvorrat* (Schütz/Luckmann 1991: 149-153) may be too eagerly glossing over its (spatial) asymmetry (see relevant criticism: de Certeau 1980: 246-249).

rical "locations" (rather than "moments"): (1) a unique living system X positioned within reach of its "private" knowledge, and (2) a *sender/recipient* pair placed in a middle of nowhere and endowed with an average cognitive background. In principle, all the observer needs to generate information is to trade a chunk of his or her "privileged access" to own states, beliefs and other unshared knowledge for the advantage of understanding (as a recipient) or being understood (as a sender): the patient answering doctor's query "How do you feel?" by telling *all* about his or her feelings, may end up getting no help at all. However, interference in a non-sequential environment is a trickier thing that one might suppose, since the multi-directionality of "spatial" asymmetry complicates its generalizing consolidation: whereas the measured flow of turn-taking will (in most cases) drag the new information, wrapped up in noise, along simply as the time of interaction goes by,[31] the topologically "symmetrical" semantics of natural language, money, law, morals and other norms and institutions will never be reaching all dark corners of "private" meanings unless all individual identities are surrendered to the homogeneous "mass ornament" (Krakauer 1927; see also Postoutenko 2009a). As a result, the privileged access remains a constant mechanism of overproducing unshared knowledge in each and every reflective being: more often than not, it transgresses the biologically necessary minimum and encroaches at potentially contested (social) areas of environment, and it stands to reason that the invaded areas may serve as convenient springboards for attacking Barbarians of all stripes. Since one of the goals of this paper is to expose the roots of asymmetrical concepts in social communication, it's worth pausing to look at presuppositions of this transgression at a close range.

The fact that the system's initial awareness of its own states is overwhelmingly asymmetrical in relation to social environment (which, at that stage, includes other systems), is almost too trivial to mention: even scholars equipped with the best instruments are still working on decoding messages carried out by motor and sensory neurons up and down the nervous system. As long as such messages remain random parts of systemic self-organization rather than its replicas, they cannot get detached, externalized, broken down into *signifi-*

31 This might be the simplest example of "active" and "constructive" role of time in politics (Rosanvallon 2002: 32; see also Palonen 2006).

er/signified, communicated in the social world and verified by others, and the difference between the system *x*'s knowledge about itself and the system *y*'s knowledge about the system *x* is complete. But this extreme degree of informational disequilibrium quickly changes with the very beginning of interaction between *x* and *y*. Even if the Dr. Jekyll's message is inadvertent and deeply personal at that (face writhed in pain), Mr. Hyde will invariably break it down, semiotically speaking, into *index* (the physical presence of his counterpart in *some* state signaling the incontestable authenticity of Dr. Jekyll's identity *here* and *now* (Tugendhat 1979: 87)) and *icon* (a reference to a visual sign—a generalized expression of a pain that may well be contested if Dr. Jekyll is prone to feign his sufferings). This relatively complex operation is indispensable if Dr. Jekyll and Mr. Hyde are to engage in communication and become a single observer, since the actual existence of sender *and* the content of his message, no matter how inextricably tied they are, must be checked separately. Most importantly, both verifications rest on postulating spatial symmetry of environment (in the first case) and the symbolic system (in the second case): the fact that Dr. Jekyll is pained but not deceased may only be ascertained against the backdrop of his physical background (taken to be immovable), and reading his face expression presupposes semantic affinity, sufficient to generate an inter-systemic interference, between references to "pain" from the "inside" (Dr. Jekyll) and from the "outside" (Mr. Hyde) position (Postoutenko 2010a: 284). It goes without saying that the second verification, in particular, is a big stretch (see Wittgenstein's saying above), but trimming Dr. Jekyll's extensive private knowledge of his inner self to the locationally unspecific—i.e. symmetrical in relation to space—social norm of "pain" is the observer's price to pay for generating information rather than noise. Dr. Jekyll may grumble at the lack of compassion, but he has no interactional resources to convert his expansive private knowledge into significant informational advantage because of its low potential value (Mr. Hyde, or anyone else, is also capable of feeling pain) and due to the constraints of one-dimensional semantics, inseparable from the speaker's systemic existence and therefore ill-suited for normative (symmetrical) conservation.

Things change, however, when Dr. Jekyll's troubles are expressed in symbolic code ('*I feel pain*'). Although the basic interplay of indexical and symbolical meanings in the utterance remains unchanged, the number of asymmetrical layers in its semantics, reconciled by pseudo-symmetries of language and other norms, increases twofold. Even the most asymmetrical part of the utterance—the first-person pronoun singular, which meaning requires from the sender simultaneous acting (as a separate living being) *and* referring (to this action), cannot be understood unless the symbol used for this reference is equally good for all speakers in similar situations independently from their locations. Consequently, the utterance '*I feel pain*' contains at least two asymmetries. The *first difference* separates sender's and recipient's states at the moment(s) of sending and receiving a message. The *second*, in turn, differentiates their locations in relation to the state described in the message, which—to the extent sufficient for keeping communication intact—are by and large reconciled by symmetries derived from background social knowledge ('*all living beings may experience pain*') and linguistic competence as its specific compartment ('*"I" means speaker, and speaker only, at the moment of speaking*').[32] Again, as the words "by and large" indicate, the cap on personal knowledge which may be included in the message leaves large pockets of potential information on the sender's side unprocessed (Luhmann 1987: 15), but normative and institutional mechanisms of natural language are usually vigorous enough to ensure such tentative reconciliations in face-to-face communication.

It is in the case of distal (mediated) communication—such as writing/reading—that the normative ceiling of interpersonal understanding may start cracking: as long as the symmetry of self-referential asymmetries cannot be confirmed on the spot, an even stronger independence of linguistic semantics from locational variance between the speaker and the recipient might be indispensable to get message across. However, the means of achieving this independence are not gradual but radical, as might be illustrated by the utterance '*I am N from M who has a fever of 38°C*'. The form of logical equation '$X \leftrightarrow Y$', perfectly symmetrical due to its *a priori* invariance to the heterogeneity of physical and social world, is filled with equally irreflexive content pertaining to the sender and affixed to its indexical self-reference as a sort of universal quality mark, aimed at certifying the validity of the statement in all parts of the (literate) world.[33] It comes

32 With some reservations, the same applies to the words "feel" and "pain".

33 Naturally, the *reading* of such an equation destroys its symmetry *ipso facto*.

as no surprise that the incongruous attempt to fuse together indexical self-performance (*'I am N'*) and its attempted translation into propositional terms (*N is Y [a human being located in M and having temperature t°]*) has dubious logical status (Heckmann 1991: 68; Tugendhat 1979: 83). However, even the relations in the predicative part of the sentence, conceived in impeccably intrinsic terms, are far from being symmetrical for various reasons, including the observer's optics (Aristotelian logic is not shared by all people in all places), the grammar of predicative statements (N is not the only sick person in the world) and the semantics of social constructs (*'N from M'* may be treated as malingerer in P where fever is not recognized as a symptom of illness). Nevertheless, the fusion of linguistic and logical norms regulating the bulk of verbal communication is strong enough for transmitting the meaning of the saying in question between two arbitrarily chosen locations without really taking their difference into account: for making sure that a signified of *'I'* is a living system, defining its permanent location (*'N from M'*) and accessing its state (*'a fever of 38°C'*), neither sender's actual distance from the recipient nor the privileged access resulting from this location matter. This stunning indulgence of the communicative system to its own semiotic inconsistency and semantic intransigence could probably be explained by independence of its regulative mechanisms from the propositional content of language. The regulative (pseudo-)symmetries—such as the convention that attributes social identity to each individual capable of reflexive self-expression (Schütz/Luckmann 1991: 93)—hover in the heights from which the subtleties of linguistic semantics are barely discernible.[34] The lower-level norms, in their turn, generally shoulder the responsibility for "objectifying" private knowledge upon interaction participants. Indeed, whereas the sender's existence as a living being is automatically confirmed in the utterance in question by the very *fact* of communicative performance rather than its verbal *content* (Postoutenko 2010a: 294), the meaning of the whole sentence depends for the most part on its recipient's willingness either to take N's saying at a

34 Despite its ethereal flair, the equation between 'I' and social identity of its utterer is neither time-nor space-symmetrical: such influential thinkers as Parmenides or St. Augustine questioned reflexivity of the first-person pronoun singular, considering Deity to be its main denotate in human communication (see Riedel 1988: 32, 47).

face value or *externally* verify N's whereabouts and temperature (Davidson 2001: 6). Still, regardless of this decision and the test outcomes, no significant results except for correspondence (or non-correspondence) between the tester's understanding of the sentence and his interpretation of the experimental data could be obtained: in other words, no light would be shed on whether N is a visually impaired person unable to read his thermometer, a compulsive liar who accidentally guessed it right, an amnesiac, a hypochondriac, or something in-between. All in all, there is little doubt that the unique degree of spatial invariance, achieved in symbolic system of verbal language through progressive intra-systemic differentiation between the observer (sender/recipient) and the thing observed (a message) (Luhmann 1987: 195), has pressed hard but by no means eliminated the asymmetrical private knowledge, which deictic expression, for all its communicative importance (see Greenberg 1966; Perkins 1992), is in reality little more than the tip of the iceberg.

Given these circumstances, the disarmingly feeble pragmatic significance of spatial symmetry in natural language could hardly raise eyebrows: unless a propositional statement, framed by sender in thoroughly intrinsic terms, is verified by the recipient from a spot not dissimilar to the sender's location, the amount of knowledge obtained by privileged access and smuggled into an "objective form" of predicative statement such as '*The cat is on the mat*', is impossible to determine. Up until now, the discussion of this inalienable spatial asymmetry in communication has been mostly focused upon biologically predetermined mechanisms of systemic self-organization in humans responsible for forging individual social identity. However, there are sufficient reasons to believe that such asymmetries are not limited to the minimum necessary for survival: rather, much to the dismay of believers in "consensus" regulating human existence (see the major programmatic text: Grice 1989, 24-30), they pop up in communication every time when the locational difference has a potential of being converted into social advantage (Harris 1995: 118-119; Bailey 1991: 476-477; Newmeyer 1990: 271; Illing 1985: 45; Coupland/Wieman/Giles 1991: 2; Ladegaard 2009: 651).

A typical example of this conversion might be spatial asymmetries in economics. Here the hopes to make up for locational disparity, pinned on normative semantics, are understandably low: it is rather uncommon to take advertisement at a face value. Furthermore, the privileged access of the producer (or seller) to the own product (or good) discredits the sentence '*This car is like new*' from the start, regardless of its being true or false. In their extreme, both asymmetries appear to pave the way to total semantic inflation *à la* Gresham and Nietzsche and spell the doom of market economy (Akerlof 1970: 489; Molho 1997: 1-2; Schwalbe 1999: 49-50). Understandably, the relative weakness of moral and linguistic norms confronts the recipient with the alternative of fighting his or her way to the position enabling privileged access (for instance, conducting independent car inspection) and relying on norms which are stronger than natural language semantics and cover areas wide enough to be considered "universal". Although the choices are usually made in favor of a less costly option (Lehmann 1999: 9-10; Campbell 2006: 4), the tendency of resorting to high-level, institutionally anchored (pseudo-)symmetries when low-level conventional ones fail, observed earlier in analysis of temporal symmetries, remains in force. In this sense, the consumer law would in theory override guarantees and warranties, although in the latter case the costs associated with referring to legal norms (hiring intermediates etc.) may well be prohibitive (Illing 1985: 42-43).

It should be noted, however, that the originally bilateral structure of spatial asymmetries, based on imaginary "equity" of irreducible knowledge deficits on both ends of communication, persists in virtually all interactional contexts, large and small, albeit in forms fairly remote from the original. In this sense, an adversary trial, where both antagonists possess nontransparent private knowledge, could be regarded as a more probable replica of spatial asymmetry *in toto* than a used car shop. (To be sure, the bench is a less liberal—or less cynical—environment than any market, so both positive (sworn) and negative (penalty of perjury) norms pressure antagonists to extend—at their costs, of course—the compulsory semantic warranties on veracity of their utterances (Harris 1995: 125-126)). However, as the role of the privileged access to own state in complex social disequilibria is naturally limited, the default presumption of this primary equity gets overshadowed by the explosive growth of asymmetries resulting from functional specializations and hierarchical distributions. Expert knowledge, state secrets (kept "above") or first-hand experience with environment (obtained "below") turn out to be much more powerful social identifiers than details of personal well-being, and the reduction of this heterogeneously asymmetrical dependencies to common (symmetrical) denominators becomes a daunting task that keeps preoccupying the minds of theoreticians and practitioners of public interaction from Plato to our days (Simmel 1907; Elias 1976: 314; Dahl 1957; Bogner 2003: 171; Hall 1985: 171; Luckenbill 1979: 107; Rubin/Zartman 1995: 359). Apparently, the inventory of such denominators cannot be limited to the most general norms of society such as law, money and verbal language, and it is bound to include at least some autopoetic replicas of interactional process *per se* such as conversation (Fumaroli 1986: 113-210). In reality, such conversational arrangements as negotiations, explicitly idolizing the *final* symmetry of positions as opposed to their original chaotic randomness, do play an important role in fabricating informational equilibria (treaties) under conditions of mutual distrust and inequality, since the defiance of procedure (regardless of motives) tends to decrease social standing in *any* form of turn taking, and negotiations are no exception from this tendency (Rubin/Zartman 1995: 359). However, the high-level normative regulation of communicative frameworks, mimicking power inequalities (Habermas 1974: 264; Miehaus 2003: 74; Bourdieu 1991: 206), frequently makes a travesty of mid-level symmetry between interchangeable conversation partners. In social systems with less developed checks and balances (see Ruesch/Bateson 1951), the ritualistic pre-allocation of turns may grant the bulk of conversation time to a single speaker. Under such conditions, the flood of speaker-centered deictic terms (such as '*I*' or '*there*'), as well as references to personal feelings ('*disturbing*') and judgments ('*erroneous*') may run unopposed: for instance, the data collected in an empirical study (Postoutenko 2011) hints at the negative correlation between the totalitarian tendencies in a given country and the number of deontic asymmetries in its political discourse. In general, the compartmentalized interactional setting (delayed response in mass communication), exacerbated by inadequate communicative means (the choice between applauds, whistling and prefabricated slogans at public gatherings) nullify the recipient's chances to produce passable interference (Cohen-Seat 1960): the staged "spontaneity" of public gatherings and their medial reproductions in Soviet Union, Nazi Germany and Fascist Italy provide ample evidence of communicative environment reinforcing, rather than taming, social asymmetries (Postoutenko 2010b: 24-27).

A somewhat subtler expropriation of recipient's communicative identity consists of presenting interference between two parts of the observer as given (past) and identical to the speaker's present position. A case in point is a catchy slogan '*This time, we are all in the front line*', uttered by George VI in his Christmas broadcast of 1940 (see Freedman 1999: 57). The lack of coordination between the propositional (plural) and the self-referential (singular) semantics, which is evident in the meaning of '*We*', and which always plays in the hands of the current speaker (Wales 1980: 33; Tietz 2002: 55), allowed His Majesty to drive "all" his real and potential audiences together at the imaginary "line" next to him: everyone else, in effect, was proclaimed nonexistent, i.e. not living in "this" time arbitrarily defined by the speaker.[35] Last but not least, the smuggling of lexical references to private emotions and evaluations into impersonal syntax of predicative statements appears to be the most peaceful way of depriving recipients of their positions. A typical example could be found in W. R. Churchill's tribute to the rescuers of London, where the seemingly disinterested characterization of British mood teems with his personal evaluations of '*wisdom*', '*rightness*', '*depth*' and '*maturity*': "The mood of Britain is wisely and rightly averse from every form of shallow or premature exultation" (Churchill 1941: 82).

All these examples hint at difficulties that Reinhart Koselleck's binary scheme may encounter in the realm of complex dissipative systems. Depriving Barbarians of the right to self-expression is an easy catch when the costs of ignoring their systemic nature are bearable: in order to achieve victory over pathogens, an immune system does not have to ascribe them to a "counter-system". In communication, however, a body can neither speak to an ear no listen to a mouth, so the relative "symmetry of asymmetries" (Luhmann 1995: 139) between '*positive HERE*' and '*negative THERE*', existing *both* in Self and Other(s), has to be recognized and dealt with (see, for instance, Paget 1967). This mutual recognition explains the fact that the loud destruction of competitors' (non-)identity, prioritized by traditional *Begriffsgeschichte*, may not be as widespread in public communication as its quiet absorption: at any rate, disguising personal emotions and egocentric evaluations as normative semantics seems to be no less a popular interaction strategy than its mirror reflection—an explicit concern

35 Four centuries before the British monarch, Martin Luther performed the same trick of imaginary integration in his inventive translation of the Apostles *Wir glauben alle an einen Gott* (Luther 1525).

with others' "abnormality" (Postoutenko 2011). But the ultimate symmetry of (spatial) asymmetries usually makes itself visible in both versions, positive and negative: simply speaking, "Hellenes", whoever they are, unmask themselves as *Barbarians*^{2} when they try to sell themselves as *Hellenes* to *Barbarians*^{1} (see the skeptical remark of Claude Levi-Strauss in the epigraph).[36] The sober look at ubiquity of barbarism and futility of semantic universalism might be depressing at times, but its grimness may save scholars a lot of trouble.

36 Particularly the historians working on the last century, second Levi-Strauss's opinion (see Hobsbawm 1994).

Figure 1

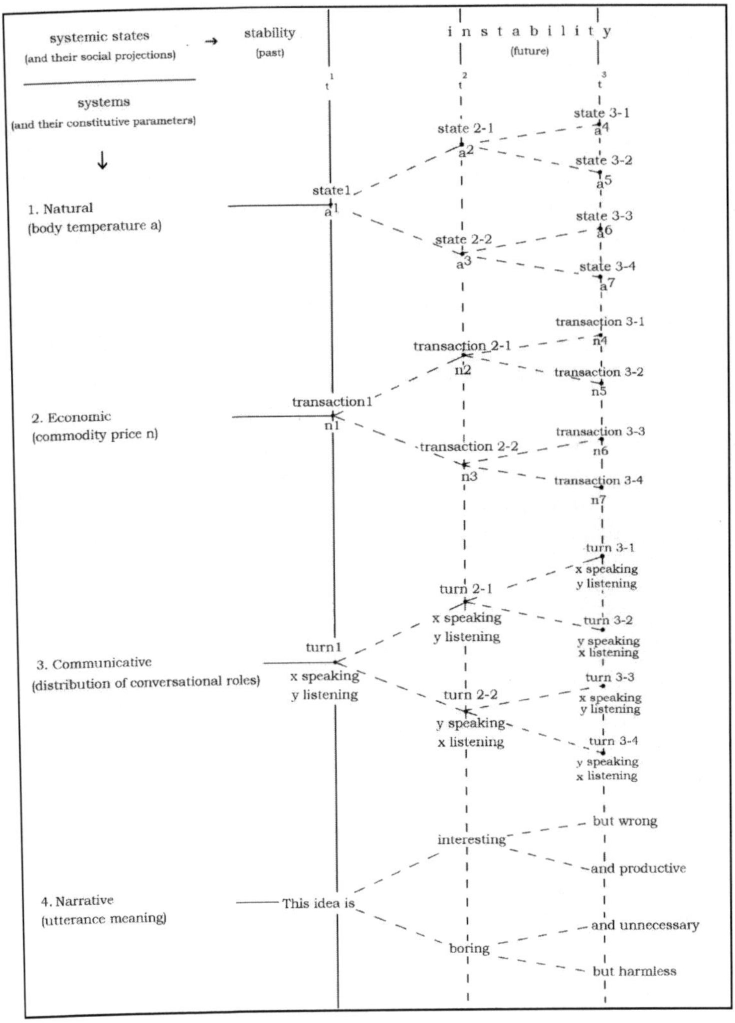

Source: Figure by Irina Valkova

REFERENCES

Adorno, Th. W. [1956]. 1990. *Zur Metakritik der Erkenntnistheorie*. Frankfurt am Main: Suhrkamp.

Adorno, Th. W. 1966. *Negative Dialektik*. Frankfurt am Main: Suhrkamp.

Akerlof, G. A. 1970. The Market for "Lemons": Quality Uncertainty and the Market Mechanism. *The Quarterly Journal of Economics*, 84 (3): 488-500.

Anspach, M. R./Varela, F. J. 1994. The Body Thinks. The Immune System in the Process of Somatic Individuation. In *Materialities of Communication*, edited by H. U. Gumbrecht/K. L. Pfeffer. Stanford: Stanford University Press, 273-285.

Arnheim, R. 1992. *To the Rescue of Art. Twenty-Six Essays*. Berkeley & Los Angeles: University of California Press.

Auguet, R. 1977. *Le juif errant: Genése d'une légende*. Paris: Payot.

Bailey, F.G. 1991. Why is Information Asymmetrical? Symbolic Behavior in Formal Organizations. *Rationality and Society* 3: 475-495.

Bataille, G. [1933] 1970 La notion de dépense. In *Œuvres complètes*, by G. Bataille. Paris: Gallimard, 302-320.

Bateson, G. 1980. *Mind and Nature. A Necessary Unity*. Glasgow: Fontana/Collins.

Bateson, G/Mead, M. 1976. [Interview]. *CoEvolutionary Quarterly* 10: 32-44.

Bauer, J. 2008. *Das kooperative Gen. Abschied vom Darwinismus*. Hamburg: Hoffmann und Campe.

Beck, D. 2001. Person-Hierarchies and the Origin of Asymmetries in Totonac Verbal Paradigms. *Linguistica Atlantica* 23: 35-68.

Bellamy, E. 1888. *Looking Backward. 2000-1887*. Boston: Houghton Mifflin Company.

Benjamin, W. [1935] 1991. Das Kunstwerk im Zeitalter seiner technischen Reproduzierbarkeit. In *Abhandlungen (Gesammelte Schriften. Band I-2)*, by W. Benjamin. Frankfurt am Main: Suhrkamp, 431-508.

Bergson, H. [1889] 1930. *Essai sur les données immédiates de la conscience*. Paris: Alcan.

Black, A. H. 1989. *Man and Nature in the Philosophical Thought of Wang Fu-chih*. Seattle: University of Washington Press.

Bogner, A. 2003. Macht und die Genese sozialer Gruppen. *Sociologus* 53 (2): 167-181.

Bourdieu, P. 1982. *Ce que parler veut dire. L'économie des echanges linguistiques.* Paris: Fayard.

Bourdieu, P. 1991. *Language et pouvoir symbolique*. Paris: Fayard.

Bradshaw, J. L. 1993. *The Evolution of Lateral Asymmetries, Language, Tool Use, and Intellect.* San Diego: Academic Press.

Bremond, C. 1973. *Logique du récit.* Paris: Seuil.

Burrow J. W. 2000. *The Crisis of Reason. European Thought, 1848-1914.* New Haven & London: Yale University Press.

Campbell, D. 2006. *Incentives. Motivation and the Economics of Information.* Cambridge: Cambridge University Press.

Certeau, M. de [1980] 1990. *L'invention du quotidien.* 1. *Arts de faire.* Paris: Gallimard.

Chaitin, G. J. 2007. *Thinking about Gödel and Turing. Essays on Complexity*. Singapore: World Scientific Publishing.

Cherry, C. 1966. *On Human Communication.* Cambridge: Cambridge University Press.

Chilton, P. 2004. *Analyzing Political Discourse.* London: Routledge.

Churchill, W. R. [1941] 1952. A Speech to the House of Commons (September 9, 1941). In *War Speeches. From June 25, 1941 to September 6, 1943*, by W. R. Churchill. London: Cassell & Company, 70-82.

Levi-Strauss, C. 1952. *Race et histoire*. Paris: UNESCO.

Cohen-Seat, G. 1960. I problemi sociologici die grandi mezzi di communicazione. *Rassegna Italiana di Sociologia* 1 (1): 45-58.

Cohn, N. 1970. *The Pursuit of the Millennium. Revolutionary Millenarians and Mystical Anarchists of the Middle Ages*. New York: Oxford University Press.

Coleridge, S. T. [1817] 1834. *Biographia Literaria; or Biographical Sketches of My Literary Life and Opinions.* New York: Leavitt, Lord & Co.

Condorcet, J.-A.-N. C., marquis de [1783] 1994. Refléxions sur la méthode de déterminer la probabilité des événements futurs d'après l'observation des événements passès. In *Arithmétique politique. Textes rares ou inédits*, by J.-A.-N. C. marquis de Condorcet. Paris: A l'Institut national d'études démographiques, 419-431.

Condorcet, J.-A.-N. C., marquis de [1794] 1988. *Esquisse d'un tableau historique des progrès de l'esprit humain suivi de Fragment sur l'Atlantide*. Paris: Flammarion.

Conrad, P/Barker, K. K. 2010. The Social Construction of Illness. Key Insights and Policy Implications. *Journal of Health and Social Behavior* 51: 67-79.

Coupland, N./Wiemann, J. M./Giles, H. 1991. Talk as "Problem" and Communication as "Miscommunication". An Integrative Analysis. In *"Miscommunication" and Problematic Talk*, edited by N. Coupland/H. Giles/J. M. Wiemann. Newbury Park: Sage, 1-17.

Cutnell, J. D./Johnson, K. W. 2005. *Physics*. New York: John Wiley and Sons.

Červenka, M. 1996. Čtyři dimenze literárního díla. In *Obléhání zevnitř*, by M. Červenka. Praha: Torst, 26-40.

Dahl, R. A. 1957. The Concept of Power. *Behavioral Science*, 2 (3), 201-215.

Davidson, D. 2001. *Subjective, Intersubjective, Objective*. Oxford: Oxford University Press.

Davies, P. C. W. 1974. *The Physics of Time Asymmetry*. Surrey: Surrey University Press.

Delumeau, J. 2000. *History of Paradise. The Garden of Eden in Myth and Tradition*. Urbana, IL & Chicago: University of Illinois Press.

Derrida, J. 1967. *De la grammatologie*. Paris: Editions de Minuit.

Dohrn-van Rossum, G. 1996. *History of an Hour*. Chicago: University of Chicago Press.

Einstein, A. [1920] 1988. *Über die spezielle und die allgemeine Relativitätstheorie*. Braunschweig: Vieweg.

Elias, N. 1970. *Was ist Soziologie?* München: Juventa-Verlag.

Elias, N. 1976. *Über den Prozess der Zivilisation*. Frankfurt am Main: Suhrkamp.

Elias, N. 1984. *Über die Zeit*. Frankfurt am Main: Suhrkamp.

Fairhurst, G. Th. 2007. *Discoursive Leadership. In Conversation with Leadership Psychology*. London: Sage.

Fodor, J. A. 1975. *The Language of Thought*. Cambridge, MA: Harvard University Press.

Foerster, H. von. 1982. *Observing Systems*. Seaside, CA: Intersystems Publications.

Foucault, M. 1969. *L'archeologie du savoir*. Paris: Gallimard.

Frank, M. 1984. Das 'fragmentarische Universum' der Romantik. In *Fragment und Totalität*, edited by L. Dällenbach/C. L. Hart Nibbrig. Frankfurt am Main: Suhrkamp, 212-224.

Frazer, J. T. 1978. *Time as Conflict*. Basel & Stuttgart: Birkhduser.

Freedman, J. 1999. *Whistling in the Dark. Memory and Culture in Wartime London*. Lexington, KY: University Press of Kentukky.

Fumaroli, M. 1986. *Trois institutions littéraires*. Paris: Gallimard.

Gavruseva, L. 1995. Positioning and Framing: Constructing Interactional Asymmetrz in Employer-Employee Discourse. *Discourse Processes* 20: 325-345.

Giesecke, M. 2006. *Der Buchdruck in der frühen Neuzeit. Eine historische Fallstudie über die Durchsetzung neuer Informations- und Kommunikationstechnologien*. Frankfurt am Main: Suhrkamp.

Godard, O./Salles J.-M. 1991. Entre nature et société. Les jeux de l'irréversibilité dans la construction économique et sociale du champ de l'environnement. In *Les figures de l'irréversibilité en économie*, edited by R. Boyer/B. Chavance/O. Godard, Paris: Éditions de l'École des hautes études en sciences sociales, 233-272.

Goffman, E. 1971. *Relations in Public. Microstudies of Public Order*. New York: Basic Books.

Goodwin, J. 1993. *Eisenstein, Cinema and History*. Urbana-Champaign, IL: University of Illinois Press.

Greenberg, J. H. 1966. Some Universals of Grammar With Particular Reference to the Meaningful Elements. In *Universals of Language*, edited by J. H. Greenberg. Cambridge: Cambridge University Press, 73-113.

Greenberg, M. 2009. *Cornelle, Classicism and the Ruses of Symmetry*. Cambridge: Cambridge University Press.

Grice, H. P. 1989. *Studies in the Way of Words*. Cambridge, MA: Harvard University press.

Guattari, F. [1993] 1995. A propos des Machines. *Chimères* 19: 85-96.

Gurevitch, A. 1990. *Kul'tura bezmovstvuiushchego bol'shinstva*. Moskva: Iskusstvo.

Haase Svendsen, G. L./Tinggaard Svendsen, G. 2003. Social kapital og økonomisk sociologi. *Distinktion* 7: 105-118.

Habermas, J. [1974] 1984. Überlegungen zur Kommunikationspathologie. In *Vorstudien und Ergänzungen zur Theorie des kommunikativen Handelns*, by J. Habermas. Frankfurt am Main: Suhrkamp, 226-270.

Hahn, A. 1986. Contribution à la sociologie de la confession et autres formes institutionnalisées d`aveu: autothématisation et processus de civilisation. *Actes de la recherche en sciences sociales* 62/63: 54-68.

Hall, P. M. 1985. Asymmetric Relationships and Processes of Power. In *Foundations of Interpretive Sociology. Original Essays in Symbolic Interaction*, edited by H. A. Farberman/R. S. Perinbanayagam. Greenwitch, NY: JAI Press, 309-344.

Hargie, O./Saunders, Ch./Dickson, D. 1994. *Social Skills in Interpersonal Communication*. London: Routledge.

Harris, S. 1995. Pragmatics and Power. *Journal of Pragmatics* 23: 117-135.

Hausmann, D. 1998. *Causal Asymmetries*. Cambridge: Cambridge University Press.

Heckmann, H. D. 1991. Wer (oder was) ich bin? Zur Deutung des intentionalen Selbstbezuges aus der Perspective der ersten Person singularis. In *Dimensionen der Selbst: Selbstbewusstsein, Reflexivitat und die Bedingungen der Kommunikation*, edited by B. Kienzle/H. Pape. Frankfurt am Main: Suhrkamp, 64-84.

Hegel, G. W. F. [1807] 1986. *Phänomenologie des Geistes*. Frankfurt am Main: Suhrkamp.

Heitland, W. E. 1909. *The Roman Republic. Volume 1*. Cambridge: Cambridge University Press.

Hellal, A. A./Heddaya, A. A./Bhargava, B. B. 1996. *Replication Techniques in Distributed Systems*. Boston: Kluwer.

Hollis Wickman, H/Korley J. N. 1998. Colloid Crystal Self-Organization and Dynamics at the Air/Water Interface. *Nature* 393: 445-447.

Holmes, J. H. 1924. *In primitive New Guinea. An account of a quarter of a century spent amongst the primitive Ipi & Namau groups of tribes of the Gulf of Papua, with an interesting description of their manner of living, their customs & habits, feasts & festivals, totems & cults*. New York: G. P. Putnam's sons.

Horwitch, P. 1987. *Asymmetries in Time*. Cambridge, MA: MIT Press.

Hobsbawm, E. 1997. Barbarism. A User's Guide. In *On History*, by E. Hobsbawm. New York: The New Press, 253-265.

Huelsenbeck, R. [1918] 2002. Dadaistisches Manifest. In *Dada Berlin. Texte. Manifeste, Aktionen*, edited by K. Riha. Stuttgart: Reclam, 22-25.

Husserl, E. [1928] 1980. *Vorlesungen zur Phaenomenologie des inneren Zeitbewusstseins*. Tübingen: Niemeyer.

Iampolski 1987. K simvolike vodopada. *Semiotika* 21, 26-41.

Illing, G. 1985. *Geld und asymmetrische Information*. Berlin: Springer.

Innis, H. 1950. *Empire and Communications*. Oxford: Oxford University Press.

Insel, P. M/Roth W. T./Price K. 2002. *Core Concepts in Health*. Boston: McGraw Hill.

Iser, W. 1993. *Das Fiktive und Das Imaginäre. Perspektiven literarischen Anthropologie*. Frankfurt am Main: Suhrkamp.

Ishida, Y. 1996. The Immune System as a Self-Identification Process. A Survey and a Proposal In: *Proceedings of the ICMAS Workshop on Immunity-Based Systems*, edited by Y. Ishida. Nara: Nara Institute of Science and Technology, 2-12.

Ivanov, V. V. 1974. On Antisymmetrical and Asymmetrical Relations in Natural Languages and Other Semiotic Systems. *Linguistics* 119: 35-40.

Ivanov, V. V. 1983. *Gerade und Ungerade. Die Asymmetrie des Gehirns und der Zeichensysteme*. Stuttgart: Hirzel.

Jacob, F. 1982. *The Possible and the Actual*. Seattle & London: University of Washington Press.

Jakobson, R. 1960. Linguistics and Poetics. In *Style in Language*, edited by T. A. Sebeok. Cambridge, MA: MIT Press, 350-377.

James, W. [1905] 1971. How Two Minds Can Know One Thing. In *Essays in Radical Empiricism*, by W. James. New York: Harper and Row, 61-67.

Jankélévitch, V. 1974. *L'irréversible et la nostalgie*. Paris: Flammarion.

Jantsch, E. 1980. The Unifying Paradigm Behind Autopoesis. Dissipative Structures. Hyper- and Ultracycles. In *Autopoiesis, Dissipative Structures, and Spontaneous Social Orders*, edited by M. Zeleny. Boulder, CO: Westview Press, 81-87.

Jauss, H.-R. 1986. *Zeit und Erinnerung in Marcel Prousts "A la recherche du temps perdu". Ein Beitrag zur Theorie des Romans*. Frankfurt am Main: Suhrkamp.

Johnston, A. 1995. El tiempo como un salmo en san Agustín (Conf. 11). *Augustinus* 40: 153-158.

Koselleck, R. [1975] 1989. Zur historisch-politischen Semantik asymmetrischer Gegenbegriffe. In *Vergangene Zukunft. Zur Semantik geschichtlicher Zeiten*, by R. Koselleck. Frankfurt am Main: Suhrkamp, 211-259.

Koselleck, R. [1994] 2010. Feindbegriffe. In *Begriffsgeschichten. Studien zur Semantik und Pragmatik der politischen und sozialen Sprache*, by R. Koselleck. Frankfurt am Main: Suhrkamp, 274-283.

Koselleck, R. 2003. *Zeitschichten. Studien zu Historik.* Frankfurt am Main: Suhrkamp.

Krakauer, S. [1927] 1977. Das Ornament der Masse. In *Das Ornament der Masse*, by S. Krakauer, Frankfurt am Main: Suhrkamp, 50-63.

Ladegaard, H. J. 2009. Pragmatic Cooperation Revisited. Resistance and Non-Cooperation as a Discursive Strategy in Asymmetrical Discourses. *Journal of Pragmatics* 42: 649-666.

Lehmann, E. 1999. *Asymmetrische Information und Werbung*. Wiesbaden: Deutscher Universitäts-Verlag.

Lewes, D. (editor). 2006. *Auto-Poetica. Representations of the Creative Process in Nineteenth-Century British and American Fiction.* Lanham, MD: Lexington Books.

Leydesdorff, L. 2006. The Biological metaphor of a Second-Order Observer and the Sociological Discourse. *Kybernetes* 35 (3/4): 531-546.

Loewenstein, K. 1973. *The Governance of Rome*. The Hague: Martinus Nijhoff.

Löwith, K. 1949. *Meaning in History*. Chicago & London: University of Chicago Press.

Luhmann, N. 1987. *Soziale Systeme. Grundriß einer allgemeinen Theorie*. Frankfurt am Main: Suhrkamp.

Luhmann, N. 1995. Jenseits von Barbarei. In *Gesellschaftsstruktur und Semantik. Studien zur Wissenssoziologie der modernen Gesellschaft. Bd. 4*, by N. Luhmann. Frankfurt am Main: Suhrkamp, 138-150.

Luckenbill, D. 1979. Power. A Conceptual Framework. *Symbolic Interaction* 2: 97-114.

Luther, M [1525] 1917. Wie glauben all an einen Gott. In *Werke. Band 1*, by M. Luther. Leipzig: Bibliographisches Institut, 14.

Luther, M. [1530] 1983. Ein Sendbrieff D. M. Lutthers. Von Dolmetzschen vnd Fürbitder heiligenn. In *Lieder e prose,* edited by E. Bonfatti. Milano: Mondadori, 248-281.

MacNeilage, P. F. 1991. The "Postural Origins Theory" of Primate Neurobiological Asymmetries. In *Biological and Behavioral Determinants of Language Development*, edited by H. Krasnegor/D. M. Rumbaugh/R. L. Schiefelbusch/M. Studdert-Kennedy. Hillsdale, NJ: Lawrence Erlbaum Associates, 165-188.

Mannheim, K. [1922] 1980. Über die Eigenart kultur-soziologischer Erkenntnis. In *Strukturen des Denkens*, by K. Mannheim. Frankfurt am Main: Suhrkamp, 33-159.

Marácz, L./Muysken, P. 1989. Introduction. In *Configurationality. Typology of Asymmetries*, edited by L. Marácz/P. Muysken. Dordrecht: Foris, 1-38.

Maturana, H. 1990. Ontology of Observing. The Biological Foundations of Self-Consciousness and the Physical Domain of Existence. In: *Beobachter. Konvergenz der Erkenntnistheorien?,* edited by N. Luhmann. München: Wilhem Fink Verlag, 47-117.

Maturana, H./Varela, H. J. 1980. *Autopoesis and Cognition. The Realization of the Living.* Dordrecht: Reidel.

Mauss, M. [1924] 1985. Essai sur le don. Forme et raison de l échange des les sociétés archaïques. In *Sociologie et anthropologie*, by M. Mauss, Paris: PUF, 146-279.

McManus. Ch. 2002. *Right Hand, Left Hand. The Origins of Asymmetry in Brains, Bodies, Atoms and Cultures.* London: Weidenfeld & Nicolson.

Miehaus, M. 2003. Das ideale Verhör. Ein theoretischer Klärungsversuch. *Zeitschrift für Rechtssoziologie* 24 (1): 71-93.

Molho, I. 1997. *The Economics of Information.* Oxford: Blackwell.

Morton, D. J. 1926. Evolution of Man's Erect Posture (preliminary report). *Journal of Morphology* 43(1): 147-179.

Neuliep, J. (editor). 1991. *Replication Research in Social Sciences.* Newbury Park: Sage Publications.

Neville, A. C. 1976. *Animal Asymmetry.* London: Edward Arnold.

Newmeyer, F. 1990. Speaker-Hearer Asymmetry as Factor of Language Evolution. A Functional Explanation for Formal Principles of Grammar. *Proceedings of the 16th Annual Meeting of Berkeley Linguistic Society* 1990: 241-247.

Paget, R. 1967. Le "social control", la regulation sociale et le pouvoir. *Revue Francaise de sociologie* 8 (2): 207-221.

Palonen, K. 2006. *The Struggle with Time. A Conceptual History of 'Politics' as an Activity*. Münster: LIT Verlag.

Parsons, T. 1982. *On Institutions and Social Evolution*. Chicago: University of Chicago Press.

Pasteur, L [1849] 1922. Recherches sur les propriétés spécifiques des deux acides qui composent l'acide racémique. In *Œuvres. T. 1. Dissymétrie moléculaire*, by L. Pasteur. Paris: Editeurs Masson et Cie, 86-120.

Peirce, C. S. [1903] 1955. Three Trychotomies of signs. In *Philosophical Writings*, by C. S. Peirce. New York: Dover, 101-104.

Perkins, R. D. 1992. *Deixis, grammar and culture*. Amsterdam: John Benjamins.

Pfeiffer, H. 1985. *Palindrome* in der Werbung. Rückläufiges bei Handelsnamen, Warenzeichen, Werbetexten. *Sprachwissenschaft* 10 (1): 53-58.

Pinchuk, G. 2001. *Immunology*. Boston: McGraw-Hill.

Poliakov, L. 1997. *Elders of Zion, Protocols of the Learned*. In *Encyclopedia Judaica*, edited by C. Roth. (CD Rom Edition). Jerusalem: Judaica Multimedia.

Popa, R. 2004. *Between Necessity and Probability. Searching for the Definition and Origin of Life*. Berlin: Springer.

Popitz, H. [1961] 2006. *Soziale Normen*. Frankfurt am Main: Suhrkamp.

Postoutenko, K. 2009a. Between "I" and "We". Studying the Grammar of Social Identity in Europe (1900–1950). *Journal of Language and Politics* 8 (2): 195–222.

Postoutenko, K. 2009b. The Influence of Anxiety. Figures of Absolute Evil in French Socialists and Dostoevsky. *Cahiers Parisiennes* 5: 417-441.

Postoutenko, K. 2010a. Social Identity as a Complementarity of Performance and Proposition. In *Bezuge des Selbst. Selbstreferentielle Prozesse in philosophischen Perspektiven*, edited by E. Balsemão Pires/B. Nonnenmacher/S. Büttner-von Stülpnagell. Coimbra: Coimbra University Press, 271-298.

Postoutenko, K. 2010b. Prolegomena to the Theory of Totalitarian Communication. In *Totalitarian Communication. Hierarchies,*

Codes and Messages, edited by K. Postoutenko. Bielefeld: transcript Verlag, 11-40.

Postoutenko 2010c. Wandering as Circulation: Dostoevsky and Marx on the "Jewish Question". In *The Economy in Jewish History. New Perspectives on the Interrelationship between Ethnicity and Economic Life*, edited by G. Reuveni/S. Wobick-Segev. New York & Oxford: Berghahn Books, 43-61.

Postoutenko 2011. Asymmetrical Concepts and Political Asymmetries. A Comparative Glance at 20th Century Democracies and Totalitarianisms From Discursive Standpoint. *See the present volume.*

Pradeau, Th. 2009. *Les limites du soi. Immunologie et identité biologique*. Montréal: Les Presses de l'Université de Montréal.

Prigogine, I./Nicolis, G. 1980. *From Being to Becoming. Time and Complexity in Social Sciences*. San Francisco: W. H. Freeman & Company.

Propp, V. 1928. *Morfologiia skazki*. Leningrad: Akademia.

Rappaport, R. 1999. *Ritual and Religion in Making of Humanity*. Cambridge: Cambridge University Press.

Ricœur, P. 1984. *Temps et récit. 2. La configuration dans le récit de fiction*. Paris: Seuil.

Riedel, M. 1988. Grund und Abgrund der Subjektivität. In *Das neuzeitliche Ich in der Literatur des 18. und 20. Jahrhunderts*, edited by U. Füllerborn/M. Engel. München: Wilhelm Fink Verlag, 29-57.

Rosanvallon, P. 2002. *Pour une histoire conceptuelle du politique*. Paris: Seuil.

Rosenfeldt, N. E. 1991. Stalinism as a System of Communication. In *Essays on Revolutionary Culture and Stalinism*, edited by J. W. Strong. Columbus: Slavica Publishers, 138-165.

Rosenthal, J. 1998, *The Myth of Dialectics. Reinterpreting the Marx-Hegel Relation*. London: Macmillan.

Derrida, J.-J. [1755] 1985. *Discours sur l'origine et les fondements de l'inégalité parmi les hommes*. Paris: Gallimard.

Ruesch, J./Bateson, G. 1951. Communication and the System of Checks and Balances. An Anthropological Approach. In *Communication. The SocialMatrix of Psychiatry*, edited by J. Ruesch/G. Bateson. New York: Norton, 150-167.

Rougemont, D. de. 1972. *L'amour et l'Occident*. Paris: Plon.

Rubin, J. Z./Zartman, W. 1995. Asymmetrical Negotiations. Some Survey Results that May Surprise. *Negotiation Journal* 4: 349-364.

Rusek, B. 2009. Palindrom jako przykład symetrii v języku. In *Język – Styl – Gatunek*, edited by M. Kita. Katowice: Wydawnictwo Uniwersytetu Śląskiego, 187-197.

Ryan J. 1991. *The Vanishing Subject. Early Psychology and Literary Modernism*. Chicago: University of Chicago Press.

Sacks, H./Schegloff, E. A./Jefferson, G. 1974. A Simplest Systematics for the Organization of Turn-Taking for Conversation. *Language* 50: 696-735.

Sahlins, M. 1972. *Stone Age Economics*. Chicago: Aldine Transaction.

Sartre, J.-P. 1960. *Critique de la raison dialectique. Tome 1*. Paris: Gallimard.

Schegloff, E. A./Jefferson, G./Sacks, H. 1977. The Preference for Self-Correction in the Organisation of Repair in Conversation. *Language* 53: 361-382.

Schegloff, E. A. 2004. Answering the Phone. In *Conversation Analysis: Studies from the First Generation*, edited by G. H. Lerner. Amsterdam: John Benjamins, 63-108.

Schegloff, E. A. 2006. Interaction. The Infrastructure for Social Institutions, the Natural Ecological Niche for Language, and the Arena in which Culture is Enacted. In *The Roots of Human Sociality*, edited by N. J. Enfield/S. J. Levinson. Oxford: Berg Publishers, 70-96.

Scholes, R./Kellogg, R. 1966. *The Nature of Narrative*. London: Oxford University Press.

Schütz A./Luckmann, T. 1991. *Strukturen der Lebenswelt. Band 1*. Frankfurt am Main: Suhrkamp.

Sciullo, A. M. di. 2003. Introduction. Asymmetry in Grammar. In *Asymmetry in Grammar. Vol. 1 (Syntax and Semantics)*, edited by A. M. Di Sciullo, Amsterdam: John Benjamins, 1-10.

Schwalbe, U. 1999. *The Core of Economies with Asymmetric Information*. Heidelberg: Springer.

Searle, J. 1969. *Speech Acts. An Essay in the Philosophy of Language*. Cambridge: Cambridge University Press.

Shannon, C./Weaver, W. 1949. *The Mathematical Theory of Communication*. Urbana, IL: The University of Illinois Press.

Simmel, G. [1890] 1989. Über soziale Differenzierung. In *Aufsätze 1887 bis 1890. Über soziale Differenzierung. Die Probleme der*

Geschichtsphilosophie, by G. Simmel. Frankfurt am Main: Suhrkamp, 109-297.

Simmel, G. [1907] 1993. Soziologie der Über- und Unterordnung. In *Aufsätze und Abhandlungen. 1901-1908*, by G. Simmel. Frankfurt am Main: Suhrkamp, 180-257.

Smullyan, R. 1991. *Godel's Incompleteness Theorems*. Oxford: Oxford University Press.

Speas, P./Tenny, C. 2003. Configurational Properties of Point of View Roles, In *Asymmetry in Grammar. Vol. 1. Syntax and Semantics*, edited by A. M. Di Sciullo. Amsterdam: John Benjamins, 315-345.

Spence, M./Zeckhauser, R. 1971. Insurance, Information & Individual Action. *American Economic Review* 61 (2): 380-387.

Stewart, J. & A. Coutinho. 2004. The Affirmation of Self. A New Perspective on the Immune System. *Artificial Life* 10 (3): 261-276.

Stevenson, R. L. [1886] 2004. The Strange Case of Dr. Jekyll and Mr. Hyde. In *The Strange Case of Dr. Jekyll and Mr. Hyde and Other Stories*, by R. L. Stevenson. New York: Barnes and Noble, 3-78.

Stöbener, A. 2008. *Economic and Psychological Theories to Cope with Information Asymmetries*. Marburg: Metropolis.

Tietz, U. 2002. *Grenzen der 'Wir'. Eine Theorie der Gemeinschaft*. Frankfurt am Main: Suhrkamp.

Todorov, T. 1969. *Grammaire du Décaméron*. The Hague: Mouton.

Tugendhat, E. 1979. *Selbstbewusstsein und Selbstbestimmung. Sprachanalytische Interpretationen*. Frankfurt am Main: Suhrkamp.

Tugendhat, E 2003. *Egozentrizität und Mystik*. München: C.-H. Beck.

Vanderstraeten, R. 2002. Parsons, Luhmann and the Theorem of Double Contingency. *Journal of Classical Sociology* 2 (1): 77-92.

Wales, K. 1980. Exophora Re-examined. The Uses of the Personal Pronoun 'We' in Present-day English. *University of East Anglia Papers in Linguistics* 12: 22-44.

Watanabe, S. 1966. Time and the Probabilistic View of the World. In *The Voices of Time*, edited by J. T. Frazer. New York: George Braziller, 527-563.

Wittgenstein, L. [1951] 1989. Bemerkungen über die Farben. Ludwig Wittgenstein. *Werkausgabe. Band 8*. Frankfurt am Main: Suhrkamp, 11-112.

White, M. J. 1981. Fatalism and Causal Determinism. An Aristotelian Essay. *Philosophical Quarterly* 31 (124): 231-241.

Zeh, H. D. 1994. *Die Physik der Zeitrichtung*. Berlin: Springer.

Authors

Junge, Kay teaches sociology at the University of Constanz. He studied in Osnabrück, Hull (England), Berlin, and Bielefeld, received his doctoral degree in 1992 from the Justus Liebig-University, Gießen, and habilitated in 2005 at the University of Constanz. He held visiting chairs and taught at Yale University, Johann Wolfgang Goethe-University, Frankfurt, and at Bielefeld University. His main research interests are in historical and theoretical sociology. His last publication in English: "The Promise of Performance" can be found in Jeffrey Alexander et al. (editors). *The Performative Turn*, Cambridge: University Press 2003; his latest publication in German is "Zepter und Kerbholz, Macht und Geld" in Thomas Schwinn et al. (editors), *Soziale Differenzierung*, Wiesbaden: VS 2011.

Manow, Philip is professor of political science at Bremen University. He studied at Marburg and Berlin, received his PhD ('Promotion') at the Free University Berlin and his 'Habilitation' at the University of Konstanz. His research interest lies in the field of Political Economy, comparative welfare state research, the German political system, and in democratic theory. Among his recent publications are *In the King's Shadow* (2010) and *Politische Ursprungsphantasien* (2011). Previously, he held positions as the head of the research unit 'Economic governance and democratic government' at the Max-Planck-Institute for the Study of Societies, Cologne, and as a professor of political science at Konstanz and Heidelberg University.

Murphy, M. Lynne is Senior Lecturer in Linguistics and Director of English Language and Linguistics at the University of Sussex, having held posts at University of the Witwatersrand (South Africa) and Baylor University (Texas) since receiving her doctorate at the University of Illinois. Her main research interest is in lexicology, particularly the semantic relations among words, as in her 2003 monograph *Semantic Relations and the Lexicon* (Cambridge U. P.). Her interest on lexical opposition has led to publications in grammatical theory, socio-semantics, child language acquisition, and aspects of discourse, cross-culturally. Her recent publications include *Lexical Meaning* (Cambridge U. P., 2010) and *Antonymy in English* with S. Jones, C. Paradis and C. Willners (Cambridge U. P., in press).

Piazza, Roberta is a lecturer in Linguistics and English Language at the University of Sussex. Her expertise is in discourse pragmatics, stylistics and sociolinguistics. She has edited a volume in Italy (Dietro il parlato) on spoken discourse and published several articles in the Journal of Pragmatics. A monograph The Discourse of Italian Cinema and Beyond by Continuum is 2011; a co-edited volume on fictional cinema and television discourse is forthcoming by John Benjamins. She has designed a joint research project on the narrative of Jewish traumatic memory. Besides papers on film discourse (in Language and Literature 2010 and Spunti e ricerche 2008), a chapter on TV news of the Iraq war is in Haarman and Lombardo eds 2008 Evaluation and Stance in War News and a joint paper on human interest stories in a corpus of Iraqi war news is forthcoming in Journal of Pragmatics. Her work is often based on a contrastive analysis of English, American and Italian data. She participated to the international EU-funded project, IntUne, and worked on the representation of Europe in the media; a joint paper on a corpus of French, English, Italian newspapers is forthcoming in an edited volume by OUP.

Postoutenko, Kirill teaches literature, sociology and anthropology at the Smolny College of Liberal Arts and Sciences in Saint-Petersburg and is currently a Senior Fellow at the Institute of Advanced Studies in Paris. He taught at various institutions in Russia, Germany and United States of America and was a research fellow in Munich, New York, Graz/Vienna, London and Paris. Major research interests include theory and history of poetry, historical sociology of identity and social history of 20th century Europe. Book publications: *'Onegin' Text in Russian Literature* (1998, in Russian) and *Soviet Culture: Codes and Messages* (2011).

Sawilla, Jan Marco is research assistant and lecturer at University of Constance. He is author of "Antiquarianismus, Hagiographie und Historie im 17. Jahrhundert. Zum Werk der Bollandisten. Ein wissenschaftshistorischer Versuch" (Tübingen 2009) and published extensively on the the history of concepts, the history of historiography and on early modern religiosity, with a special focus on the cult of saints. His current fields of research include the history of early modern governmentality and the history of the perception of space and environment.

Vuori, Juha A. is a senior research fellow at the Department of political science and contemporary history at the University of Turku in Finland. He completed his degrees at the same department and has been teaching there since 2006. Vuori has also taught courses on East Asian politics at the University of Helsinki in Finland and Fudan University in Shanghai. He has been the editor-in-chief of Kosmopolis, the journal of the Finnish Peace Research Association, and he is the current president of the Finnish International Studies Association. His research has mainly focused on the political functions of security in the People's Republic of China, and the critical development of securitisation theory. He has co-edited and written for several edited volumes, and his articles have appeared in journals such as European Journal of International Relations, Security Dialogue, Asian Journal of Political Science, and Issues & Studies.